ORGANIC SYNTHESES

ORGANIC SYNTHESES

AN ANNUAL PUBLICATION OF SATISFACTORY
METHODS FOR THE PREPARATION
OF ORGANIC CHEMICALS
VOLUME 79
2002

A JOHN WILEY & SONS, INC., PUBLICATION

The procedures in this text are intended for use only by persons with prior training in the field of organic chemistry. In the checking and editing of these procedures, every effort has been made to identify potentially hazardous steps and to eliminate as much as possible the handling of potentially dangerous materials; safety precautions have been inserted where appropriate. If performed with the materials and equipment specified, in careful accordance with the instructions and methods in this text, the Editors believe the procedures to be very useful tools. However, these procedures must be conducted at one's own risk. Organic Syntheses, Inc., its Editors, who act as checkers, and its Board of Directors do not warrant or guarantee the safety of individuals using these procedures and hereby disclaim any liability for any injuries or damages claimed to have resulted from or related in any way to the procedures herein.

For general information on our other products and services please contact our Customer Care Department within the U.S. at 877-762-2974, outside the U.S. at 317-572-3993 or fax 317-572-4002.

Wiley also publishes its books in a variety of electronic formats. Some content that appears in print, however, may not be available in electronic format.

"John Wiley & Sons, Inc. is pleased to publish this volume of Organic Syntheses on behalf of Organic Syntheses, Inc. Although Organic Syntheses, Inc. has assured us that each preparation contained in this volume has been checked in an independent laboratory and that any hazards that were uncovered are clearly set forth in the write-up of each preparation, John Wiley & Sons, Inc. does not warrant the preparations against any safety hazards and assumes no liability with respect to the use of the preparations."

For ordering and customer service, call 1-800-CALL-WILEY.

Library of Congress Catalog Card Number: 21-17747
ISBN 0-471-41530-8

Printed in the United States of America

10 9 8 7 6 5 4 3 2 1

ORGANIC SYNTHESES

Out of print.
† *Deceased.*

Out of print.
†*Deceased.*

Collective Volumes, Collective Indices to Collective Volumes I–VIII, Volumes 75–78, and Reaction Guide are available from John Wiley & Sons, Inc.

*Out of print.
†Deceased.

NOTICE

With Volume 62, the Editors of *Organic Syntheses* began a new presentation and distribution policy to shorten the time between submission and appearance of an accepted procedure. The soft cover edition of this volume is produced by a rapid and inexpensive process, and is sent at no charge to members of the Organic Divisions of the American and French Chemical Society, The Perkin Division of the Royal Society of Chemistry, and The Society of Synthetic Organic Chemistry, Japan. The soft cover edition is intended as the personal copy of the owner and is not for library use. A hard cover edition is published by John Wiley & Sons, Inc. in the traditional format, and differs in content primarily in the inclusion of an index. The hard cover edition is intended primarily for library collections and is available for purchase through the publisher. Annual Volumes 70–74 have been incorporated into a new five-year version of the collective volumes of *Organic Syntheses* which has appeared as *Collective Volume Nine* in the traditional hard cover format. It is available for purchase from the publishers. The Editors hope that the new *Collective Volume* series, appearing twice as frequently as the previous decennial volumes, will provide a permanent and timely edition of the procedures for personal and institutional libraries. The Editors welcome comments and suggestions from users concerning the new editions.

Organic Syntheses has joined the age of electronic publication with the release of its website, www.orgsyn.org. This site is available free of charge to all chemists and contains all of the nine Collective as well as Annual Volumes and Indices.

To create the *Organic Syntheses* web site, the Board of Directors of *OS* formed a collaboration with CambridgeSoft Corporation (Cambridge, MA), producers of ChemOffice and ChemDraw, and DataTrace Publishing Company (Towson, MD), publishers of *ChemTracts*. *OS* fully funded this extensive effort. All of the information in the *OS* Collective Volumes, Annual Volumes, and Indices were digitized, mapped, and converted to XML documents by DataTrace. CambridgeSoft developed the website incorporating the databases linking text and chemical structures using their proprietary *ChemOffice Webserver* software. Reaction diagrams are stored in a ChemFinder database to facilitate structure-based searching.

The *OS* website goes far beyond the scope of the printed version and is fully searchable using a variety of techniques. Using the free *ChemDraw* plugin* for Netscape Navigator or Microsoft Internet Explorer, chemists can draw structural queries directly on the web page and combine structural or reaction transformation queries with full-text and bibliographic search terms, such as chemical name, reagents, molecular formula, apparatus, reagents, or even a hazard or warning phrase. The preparations are categorized into nearly 300 specific reaction types, allowing search by category.

*Because of browser incompatibility, at this time Macintosh users are limited to using Netscape, versions 4.5–4.73. Please read the hardware and software requirements and follow the instructions for plugin installation prior to attempting searches at the website.

NOMENCLATURE

Both common and systematic names of compounds are used throughout this volume, depending on which the Editor-in-Chief felt was more appropriate. The *Chemical Abstracts* indexing name for each title compound, if it differs from the title name, is given as a subtitle. Systematic *Chemical Abstracts* nomenclature, used in both the recent Collective Indexes for the title compound and a selection of other compounds mentioned in the procedure, is provided in an appendix at the end of each preparation. Registry numbers, which are useful in computer searching and identification, are also provided in these appendixes. Whenever two names are concurrently in use and one name is the correct *Chemical Abstracts* name, that name is preferred.

SUBMISSION OF PREPARATIONS

Organic Syntheses welcomes and encourages submission of experimental procedures which lead to compounds of wide interest or which illustrate important new developments in methodology. The Editorial Board will consider proposals in outline format as shown below, and will request full experimental details for those proposals which are of sufficient interest. Submissions which are longer than three steps from commercial sources or from existing *Organic Syntheses* procedures will be accepted only in unusual circumstances.

Organic Syntheses Proposal Format

1) Authors
2) Title
3) Literature reference or enclose preprint if available
4) Proposed sequence
5) Best current alternative(s)
6) a. Proposed scale, final product:
   b. Overall yield:
   c. Method of isolation and purification:
   d. Purity of product (%):
   e. How determined?

7) Any unusual apparatus or experimental technique?
8) Any hazards?
9) Source of starting material?
10) Utility of method or usefulness of product

Submit to: Dr. Jeremiah P. Freeman, Secretary
            Department of Chemistry
            University of Notre Dame
            Notre Dame, IN 46556

Proposals will be evaluated in outline form, again after submission of full experimental details and discussion, and, finally by checking experimental procedures. A form that details the preparation of a complete procedure (Notice to Submitters) may be obtained from the Secretary.

Additions, corrections, and improvements to the preparations previously published are welcomed; these should be directed to the Secretary. However, checking of such improvements will only be undertaken when new methodology is involved. Substantially improved procedures have been included in the Collective Volumes in place of a previously published procedure.

ACKNOWLEDGMENT

Organic Syntheses wishes to acknowledge the contributions of Discovery Partners Intl., Hoffmann-La Roche, Inc., Merck & Co., and Pfizer, Inc. to the success of this enterprise through their support, in the form of time and expenses, of members of the Boards of Directors and Editors.

HANDLING HAZARDOUS CHEMICALS
A Brief Introduction

General Reference: *Prudent Practices in the Laboratory*; National Academy Press; Washington, DC, 1995.

Physical Hazards

Fire. Avoid open flames by use of electric heaters. Limit the quantity of flammable liquids stored in the laboratory. Motors should be of the nonsparking induction type.

Explosion. Use shielding when working with explosive classes such as acetylides, azides, ozonides, and peroxides. Peroxidizable substances such as ethers and alkenes, when stored for a long time, should be tested for peroxides before use. Only sparkless "flammable storage" refrigerators should be used in laboratories.

Electric Shock. Use 3-prong grounded electrical equipment if possible.

Chemical Hazards

Because all chemicals are toxic under some conditions, and relatively few have been thoroughly tested, it is good strategy to minimize exposure to all chemicals. In practice this means having a good, properly installed hood; checking its performance periodically; using it properly; carrying out most operations in the hood; protecting the eyes; and, since many chemicals can penetrate the skin, avoiding skin contact by use of gloves and other protective clothing at all times.

a. Acute Effects. These effects occur soon after exposure. The effects include burn, inflammation, allergic responses, damage to the eyes, lungs, or nervous system (e.g., dizziness), and unconsciousness or death (as from overexposure to HCN). The effect and its cause are usually obvious and so are the methods to prevent it. They generally arise from inhalation or skin contact, so should not be a problem if one follows the admonition "work in a hood and keep chemicals off your hands". Ingestion is a rare route, being generally the result of eating in the laboratory or not washing hands before eating.

b. Chronic Effects. These effects occur after a long period of exposure or after a long latency period and may show up in any of numerous organs. Of the chronic effects of chemicals, cancer has received the most attention lately. Several dozen chemicals have been demonstrated to be carcinogenic in man and hundreds to be carcinogenic to animals. Although there is no simple correlation between carcinogenicity in animals and in man, there is little doubt that a significant proportion of the chemicals used in laboratories have some potential for carcinogenicity in man. For this and other reasons, chemists should employ good practices.

The key to safe handling of chemicals is a good, properly installed hood, and the referenced book devotes many pages to hoods and ventilation. It recommends that in a laboratory where people spend much of their time working with chemicals there should be a hood for each two people, and each should have at least 2.5 linear feet (0.75 meter) of working space at it. Hoods are more than just devices to keep undesirable vapors from the laboratory atmosphere. When closed they provide a protective barrier between chemists and chemical operations, and they are a good containment device for spills. Portable shields can be a useful supplement to hoods, or can be an alternative for hazards of limited severity, e.g., for small-scale operations with oxidizing or explosive chemicals.

Specialized equipment can minimize exposure to the hazards of laboratory operations. Impact resistant safety glasses are basic equipment and should be worn at all times. They may be supplemented by face shields or goggles for particular operations, such as pouring corrosive liquids. Because skin contact with chemicals can lead to skin irritation or sensitization or, through absorption, to effects on internal organs, protective gloves should be worn at all times.

Laboratories should have fire extinguishers and safety showers. Respirators should be available for emergencies. Emergency equipment should be kept in a central location and must be inspected periodically.

MSDS (Materials Safety Data Sheets) sheets are available from the suppliers of commercially available reagents, solvents, and other chemical materials; anyone performing an experiment should check these data sheets before initiating an experiment to learn of any specific hazards associated with the chemicals being used in that experiment.

DISPOSAL OF CHEMICAL WASTE

General Reference: *Prudent Practices in the Laboratory*, National Academy Press, Washington, D.C. 1995

Effluents from synthetic organic chemistry fall into the following categories:

1. **Gases**

   1a. Gaseous materials either used or generated in an organic reaction.
   1b. Solvent vapors generated in reactions swept with an inert gas and during solvent stripping operations.
   1c. Vapors from volatile reagents, intermediates and products.

2. **Liquids**

   2a. Waste solvents and solvent solutions of organic solids (see item 3b).
   2b. Aqueous layers from reaction work-up containing volatile organic solvents.
   2c. Aqueous waste containing non-volatile organic materials.
   2d. Aqueous waste containing inorganic materials.

3. **Solids**

   3a. Metal salts and other inorganic materials.
   3b. Organic residues (tars) and other unwanted organic materials.
   3c. Used silica gel, charcoal, filter aids, spent catalysts and the like.

The operation of industrial scale synthetic organic chemistry in an environmentally acceptable manner* requires that all these effluent categories be dealt with properly. In small scale operations in a research or academic set-

---

*An environmentally acceptable manner may be defined as being both in compliance with all relevant state and federal environmental regulations *and* in accord with the common sense and good judgement of an environmentally aware professional.

ting, provision should be made for dealing with the more environmentally offensive categories.

1a. Gaseous materials that are toxic or noxious, e.g., halogens, hydrogen halides, hydrogen sulfide, ammonia, hydrogen cyanide, phosphine, nitrogen oxides, metal carbonyls, and the like.

1c. Vapors from noxious volatile organic compounds, e.g., mercaptans, sulfides, volatile amines, acrolein, acrylates, and the like.

2a. All waste solvents and solvent solutions of organic waste.

2c. Aqueous waste containing dissolved organic material known to be toxic.

2d. Aqueous waste containing dissolved inorganic material known to be toxic, particularly compounds of metals such as arsenic, beryllium, chromium, lead, manganese, mercury, nickel, and selenium.

3. All types of solid chemical waste.

Statutory procedures for waste and effluent management take precedence over any other methods. However, for operations in which compliance with statutory regulations is exempt or inapplicable because of scale or other circumstances, the following suggestions may be helpful.

Gases

Noxious gases and vapors from volatile compounds are best dealt with at the point of generation by "scrubbing" the effluent gas. The gas being swept from a reaction set-up is led through tubing to a (large!) trap to prevent suckback and on into a sintered glass gas dispersion tube immersed in the scrubbing fluid. A bleach container can be conveniently used as a vessel for the scrubbing fluid. The nature of the effluent determines which of four common fluids should be used: dilute sulfuric acid, dilute alkali or sodium carbonate solution, laundry bleach when an oxidizing scrubber is needed, and sodium thiosulfate solution or diluted alkaline sodium borohydride when a reducing scrubber is needed. Ice should be added if an exotherm is anticipated.

Larger scale operations may require the use of a pH meter or starch/iodide test paper to ensure that the scrubbing capacity is not being exceeded.

When the operation is complete, the contents of the scrubber should be handled as aqueous waste, as outlined in the "Liquids" section that follows. In many instances, this will require neutralization, followed by concentration to a minimum volume, or concentration to dryness before disposal as concentrated liquid or solid chemical waste.

Liquids

Every laboratory should be equipped with a waste solvent container in which *all* waste organic solvents and solutions are collected. The contents of these containers should be periodically transferred to properly labeled waste solvent drums and arrangements made for contracted disposal in a regulated and licensed incineration facility.**

Aqueous waste containing dissolved toxic organic material should be decomposed *in situ*, when feasible, by adding acid, base, oxidant, or reductant. Otherwise, the material should be concentrated to a minimum volume and added to the contents of a waste solvent drum.

Aqueous waste containing dissolved toxic inorganic material should be evaporated to dryness and the residue handled as a solid chemical waste.

Solids

Soluble organic solid waste can usually be transferred into a waste solvent drum, provided near-term incineration of the contents is assured.

Inorganic solid wastes, particularly those containing toxic metals and toxic metal compounds, used Raney nickel, manganese dioxide, etc. should be placed in glass bottles or lined fiber drums, sealed, properly labeled, and arrangements made for disposal in a secure landfill.** Used mercury is particularly pernicious and small amounts should first be amalgamated with zinc or combined with excess sulfur to solidify the material.

Other types of solid laboratory waste including used silica gel and charcoal should also be packed, labeled, and sent for disposal in a secure landfill.

Special Note

Since local ordinances may vary widely from one locale to another, one should always check with appropriate authorities. Also, professional disposal services differ in their requirements for segregating and packaging waste.

**If arrangements for incineration of waste solvent and disposal of solid chemical waste by licensed contract disposal services are not in place, a list of providers of such services should be available from a state or local office of environmental protection.

CHARLES C. PRICE
July 13, 1913–February 11, 2001

Charles Coale Price, III, was born to the Quaker couple, Thornton Walton Price of Plymouth Meeting, PA, and Helen Marot Farley of Swarthmore, PA, in 1913, the first of five children. His parents had married in 1911 under the care of Swarthmore Friends Meeting, both having attended Swarthmore College. Thornton graduated from the University of Pennsylvania in 1908 as a mechanical engineer and became an entrepreneur eventually moving his family to a two hundred acre farm in northern New Jersey, where Charles grew up. At age six, Charles had his right hand blown away by an exploding box of dynamite caps. In spite of this handicap, he excelled in sports all his life. At his graduation from the George School, his father presented him an 18-foot sailboat. This was the start of a devotion to sailing, which threaded through all his other life accomplishments and forged deep bonds with family and many lifelong friends.

In 1930, he entered Swarthmore College, receiving his B.A. in chemistry with high honors and Phi Beta Kappa. He played varsity lacrosse, was captain of the team and honorable mention All-American. He met his future wife and mother of his five children, Mary Elma White, while a junior at Swarthmore. They were married on June 1936, just after her graduation from Swarthmore and his from Harvard, where he received his Ph.D. in June 1936, finishing his doctoral thesis under Louis Fieser in only twenty months. His research resulted in four papers published in the Journal of the American Chemical Society. That fall he accepted an offer to do post-doctoral work with Prof. Roger Adams at the University of Illinois, where he remained on the faculty for nine more years.

The war years were busy with many active research programs related to the war effort. Three were of prime importance. First was a major project to devise testing for known chemical warfare agents in water and to design the field equipment to remove them. Second were key studies to formulate useful substitutes for no-longer-available quinine to treat malaria in the South Pacific, which led to development of a practical industrial synthesis of chloroquin. Third were studies to understand processes important to synthetic rubber production, such as emulsion polymerization and critical means of regulating molecular weight of the synthetic rubber. His growing interest in and contributions to understanding factors affecting polymer properties and polymerization processes led to his becoming founding co-editor of the Journal of Polymer Science. This wartime research at Illinois and the consequent pool of knowledge and experience with chemical warfare agents led eventually to interesting work related to drugs for cancer.

Just before the end of World War II, Charles was invited to become the head of the Chemistry department at the University of Notre Dame, which had a good reputation in chemistry largely due to the involvement of Father Julius Nieuwland, C.S.C. with duPont in the development of the specialty rubber, Neoprene. He accepted the challenge and headed to South Bend in August 1946 with his growing family: Patricia, 8, Susanne, 6, Sarah Shoemaker, 2, and Judith Spencer, 3 months. The final family member, Charles Coale Price IV, would arrive two years later. Upon his arrival at Notre Dame, he initiated the Reaction Mechanisms Conference, a biennial event that continues to this day. At Notre Dame, Charles invented and patented polypropylene oxide-polyurethane rubber, which became the basis for most of the foam rubber produced in the world since. He was also asked to chair two national conferences on developing rubber, one for use in arctic conditions and a second for uses at high temperatures. Both conferences stimulated new approaches to practical answers for both needs. Also while at Notre Dame, Charles joined the Board of Editors of Organic Syntheses on which he served from 1946–1954. He was editor of Volume 33 of this series.

In 1947, influenced by his brother, Thornton, Charles became deeply involved with the World Federalists Association, which advocated strengthening the United Nations into a true world federal government to prevent world war through world law. Through his interest, Charles entered politics, running as the Democratic nominee for U.S. Senator from Indiana in 1950 and for the U.S. House congressional seat in the 3rd District in 1952. He became national president of the American World Federalists in 1958 and, later, chairman of the Federation of American Scientists and of the Council for Livable World. He also held leadership positions in the Commission to Study the Organization of Peace. He was convinced to his death that law is the alternative to unbridled violence as a way to organize human beings in effective, orderly ways for the common good.

In 1953, he received an offer from the University of Pennsylvania, one of the oldest and most distinguished universities in the United States, which was also located in a major center of chemical industries. In 1954, he accepted the position as the Benjamin Franklin Professor of Chemistry and chairman of the Chemistry Department. In the transitional summer of 1954, he gave a series of lectures on polymer chemistry at the Brooklyn Polytechnic Institute, which became the basis of his first book, *Reactions at the Carbon-Carbon Double Bond*. An important original idea posited in this book was the phenomenon of pi-bonding. From the beginning of his doctoral research at Harvard, a major common thread to many of Charles' contributions was his keen interest in the detailed mechanisms by which chemical reactions occur.

His years at Penn were challenging and productive. His work with polymers at elevated temperatures was later used in commercial production and the research on anti-malarial drugs and chemical warfare agents that he had done previously at Illinois during World War II was further tested on cancer patients. He also served as chair of the Faculty Senate during the height of the student Vietnam protests, was a member of an open search committee for a new President when Gaylord Harnwell resigned, and successfully chaired a committee to solicit contributions from the faculty toward a major University fund-raising drive. In spite of these extra duties, his real love was always chemistry and the amazing and crucially important "universe" of atoms and molecules that comprise the physical world around us. He felt that one of the great triumphs of human curiosity and human ingenuity has been the ability to develop instruments able to tell us about this "universe" and to teach us how to make use of that knowledge to improve the human condition. He was privileged to play a role in this great adventure.

During this time, he also served on the College Board of Managers for his alma mater, Swarthmore College, from 1954 to the mid-1970s, the last five years as Chairman. In addition, his return to Philadelphia presented many opportunities to participate in Quaker activities, such as five years as clerk of the Philadelphia Yearly Meeting working group on world federal government and Clerk of the Old Haverford Meeting. After eighteen years away from the game of lacrosse, he enjoyed playing with the Philadelphia Lacrosse Club for four years. In the fall of 1962, he spent four months in Japan with his wife and three younger children, where he taught as a Fulbright Professor at the Universities of Osaka and Kyoto.

Charles' passion for sailing started in his boyhood, racing Barnegat Bay sneak boxes, and brought him dozens of championship trophies in various classes. In 1954, he sailed his 48′ sailboat from Lake Michigan to the Chesapeake Bay where he raced extensively, winning many season trophies and participating in six Newport-to-Bermuda races (1960 to 1970). He was a longtime member of the Cruising Club of America. In 1970, he sailed his 45′ boat from Bermuda to Ireland and England, representing the CCA in the prestigious Cowes Race in which he competed against Prince Phillip and Prime Minister Ted Heath. He then went on to Munich for a five month sabbatical leave where he wrote the first draft of his last and most important book, entitled *Energy and Order, Some Reflections on Evolution (1983).* He owned many racing and cruising sailboats until, in 1993, at the age of eighty, he finally sold his last boat. However, he continued to sail, borrowing the single-handing a 37-foot ketch that belonged to a close friend and former crewmember.

In the summer of 1982, after devoting a year to her home care, he lost his wife, Mary Elma, to cancer. His marriage to Mary Elma had been such a mainstay of his life that he was committed to the institution of matrimony. Within the year he fell in love with and married Anne Parker Gill, of Swarthmore, PA, thus expanding his family to include the growing families of her two sons, Bill and Doug, to whom he devoted much time in his later years. Within the next few years, he and Anne visited the USSR, invited by the Soviet Peace Committee to promote the World Federalists Association position, and China, invited by Academia Sinica. His five children, Patricia (b. 1938), Susanne (b. 1940), Sarah Shoemaker (Sally, b. 1944), Judith Spencer (b. 1946) and Charles Coale IV (b. 1948) were a source of much pride and pleasure to him.

An activity that started at Penn, becoming his primary effort after retirement, was his role as founding Board Chair for the Chemical Heritage Foundation, an organization intended to collect and record the history of chemical sciences and industries, further chemical research and education, and enhance the public's understanding of the chemical sciences. While President of the American Chemical Society in 1965, Charles chaired their newly founded committee on Chemistry and Public Affairs. With Prof. Arnold Thackray, head of the Penn department of History and Sociology of Science, he took on the major task of finding funding, space, and staff for the Chemical Heritage

Foundation. CHF is now magnificently housed in a remodeled 150-year-old neo-classical structure at the heart of Independence National Historical Park in Philadelphia. Charles' personal interest in this venture was due in part to his strong conviction that chemistry has been the central science of the 20th century. The great accomplishment of chemists has been in revealing that long-chain molecules—polymers—can be formed into such wonders as nylon and other modern plastics and fibers, and that they form the very structure of the human genome and thus of life itself. He believed that the CHF would play a key role in helping to shape a "survivable future" for mankind and was honored that a Charles C. Price Fellowship in Polymer History was established through the Beckman Center at CHF to investigate the discoveries of the next century.

After a long bout with multiple strokes, Charles passed away peacefully in his bed at the Quadrangle in Haverford, where he lived with his wife, Anne, for the last ten years. Present with him were his son, Charles, his wife Anne Gill Price, and her son, Douglas Gill. His brother, Thornton Price, and his sisters, Jeanne Price Norman and Helen Belser, predecease him. He is survived by his youngest sister, Elizabeth Price, of Maui, his son, Charles C. Price, IV, of San Diego, CA, and his four daughters, Patricia Paxson of London, Susanne Neal of San Diego, CA, Sally Lindsay Honey of Palo Alto, CA, and Judith Price Waterman of Redwood Valley, CA.

The American Chemical Society recognized his scientific and societal interests by its Award in Pure Chemistry (1946), its Award for Creative Invention (1974), and its Parsons Award for Distinguished Public Service (1973). In 1966, the American Institute of Chemists awarded him their Chemical Pioneer Award. He was elected as president of the American Chemical Society in 1965. His ACS presidential address was on "The Synthesis of Life," which attracted considerable notoriety. It also led to a continuing interest in the origin and evolution of the universe and the solar system, and of life on planet Earth, culminating in two books: *The Synthesis of Life (1974)* and *Energy and Order: Some Reflections on Evolution* (1983). In addition to these books, he edited the text book, *Coordination Polymerization (1981)*, and wrote *Geometry of Molecules (1971)*, *Reactions at the Carbon-Carbon Double Bond (1955)*, and more than three hundred scientific papers. He was awarded more than twenty patents and fourteen honorary degrees.

He repeatedly expressed the concurrence of the two main themes of his life, one scientific and professional, the other social and political, in the following statement:

"An essential lesson for humans is to recognize the absolutely fundamental importance, throughout the sweep of evolutionary history, of ENHANCED CAPABILITIES from INCREASED ORDER by the INVESTMENT OF

ENERGY through COOPERATIVE PHENOMENA. This principle has governed the creation and function of stars, of atoms, of molecules, of living cells, of living beings, of the human brain and of human society. Humans can and must apply this principle to their evolution."

SALLY LINDSAY HONEY
Daughter

Palo Alto, CA

THEODORA W. GREENE

The Board of Editors of Organic Syntheses dedicates this volume to Theodora W. Greene upon her retirement as Assistant Editor. For 22 years Theo has edited our procedures for consistent style, has provided the Chemical Abstract names and registry numbers that appear at the end of each procedure, and has verified all the references. Quiet and unassuming, she carried out these tasks thoroughly and conscientiously; all of us in the enterprise are deeply indebted to her.

CONTENTS

ETHYL 3-(p-CYANOPHENYL)PROPIONATE FROM ETHYL 3-IODOPROPIONATE AND p-CYANOPHENYLZINC BROMIDE

Anne Eeg Jensen, Florian Kneisel, and Paul Knochel

PREPARATION OF n-BUTYL 4-CHLOROPHENYL SULFIDE

J. Christopher McWilliams, Fred J. Fleitz, Nan Zheng, and Joseph D. Armstrong III

SYNTHESIS OF SYMMETRICAL trans-STILBENES BY A DOUBLE HECK REACTION OF (ARYLAZO)AMINES WITH VINYLTRIETHOXYSILANE: trans-4,4'-DIBROMOSTILBENE

Saumitra Sengupta and Subir K. Sadhukhan

SYNTHESIS OF (+)-(1S,2R)- AND (-)-(1R,2S)-trans-2-PHENYLCYCLOHEXANOL VIA SHARPLESS ASYMMETRIC DIHYDROXYLATION (AD)

Javier Gonzalez, Christine Aurigemma, and Larry Truesdale

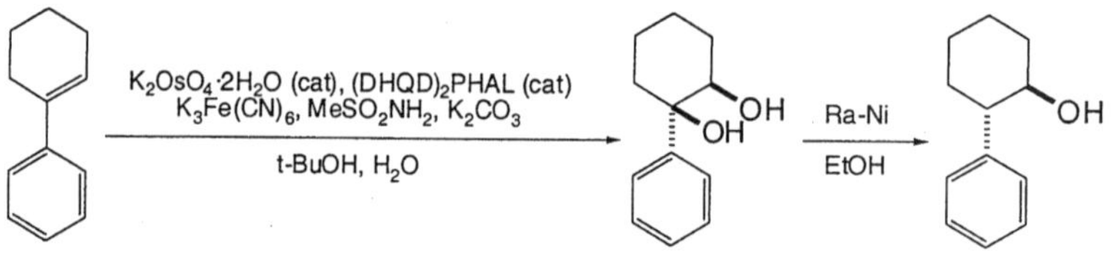

DICYCLOHEXYLBORON TRIFLUOROMETHANESULFONATE

Atushi Abiko

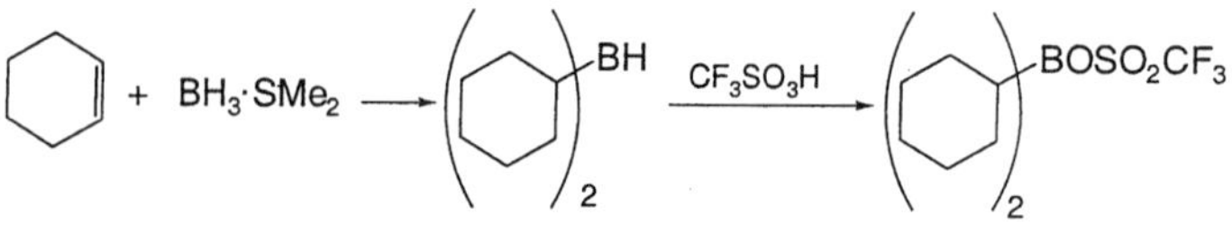

2-(N-BENZYL-N-MESITYLENESULFONYL)AMINO-1-PHENYL-1-PROPYL PROPIONATE

Atushi Abiko

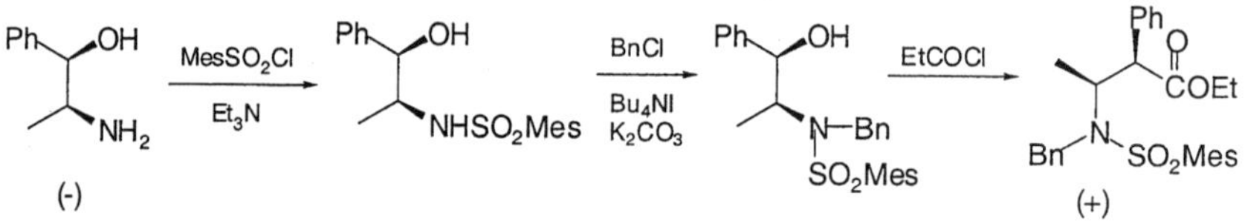

ANTI-SELECTIVE BORON-MEDIATED ASYMMETRIC ALDOL REACTION OF CARBOXYLIC ESTERS: SYNTHESIS OF (2S,3R)-2,4-DIMETHYL-1,3-PENTANEDIOL

Atushi Abiko

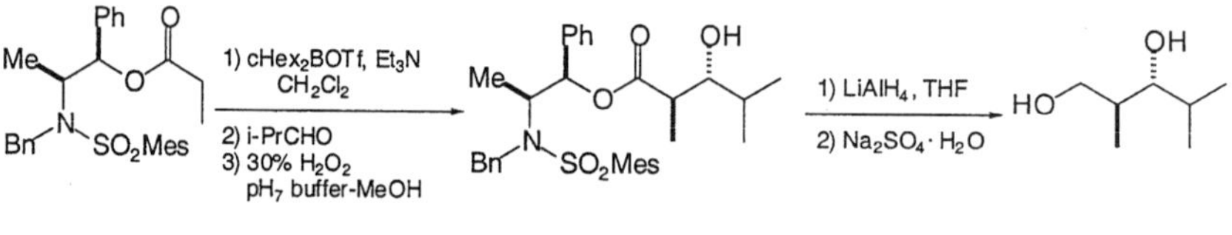

(S)-3-(tert-BUTYLOXYCARBONYLAMINO)-4-PHENYLBUTANOIC ACID

Michael R. Linder, Steffen Steurer, and Joachim Podlech

PHOTO-INDUCED RING EXPANSION OF 1-TRIISOPROPYLSILYLOXY-1-AZIDOCYCLOHEXANE: PREPARATION OF ε-CAPROCACTAM

Jade D. Nelson, Dilip P. Modi, and P. Andrew Evans

(3,4,5-TRIFLUOROPHENYL)BORONIC ACID-CATALYZED AMIDE FORMATION FROM CARBOXYLIC ACIDS AND AMINES: N-BENZYL-4-PHENYLBUTYRAMIDE

Kazuaki Ishihara, Suguru Ohara, and Hisashi Yamamoto

PREPARATION OF SECONDARY AMINES FROM PRIMARY AMINES VIA 2-NITROBENZENESULFONAMIDES: N-(4-METHOXYBENZYL)-3-PHENYLPROPYLAMINE

Wataru Kurosawa, Toshiyuki Kan, and Tohru Fukuyama

2-AMINO-3-FLUOROBENZOIC ACID

Martin Kollmar, Richard Parlitz, Stephan R. Oevers, and Günter Helmchen

3-(4-BROMOBENZOYL)PROPANOIC ACID

Alexander J. Seed, Vaishali Sonpatki, and Mark R. Herbert

4-METHOXYCARBONYL-2-METHYL-1,3-OXAZOLE

James D. White, Christian L. Kranemann, and Punlop Kuntiyong

[4+3] CYCLOADDITION IN WATER.
SYNTHESIS OF 2,4-endo, endo-DIMETHYL-8-OXABICYCLO[3.2.1]OCT-6-EN-3-ONE

Mark Lautens and Giliane Bouchain

ORGANIC SYNTHESES

SYNTHESIS OF TRIS(2-PERFLUOROHEXYLETHYL)TIN HYDRIDE:

A HIGHLY FLUORINATED TIN HYDRIDE WITH ADVANTAGEOUS FEATURES

OF EASY PURIFICATION

[Stannane, tris-(3,3,4,4,5,5,6,6,7,7,8,8,8-tridecafluorooctyl)-]

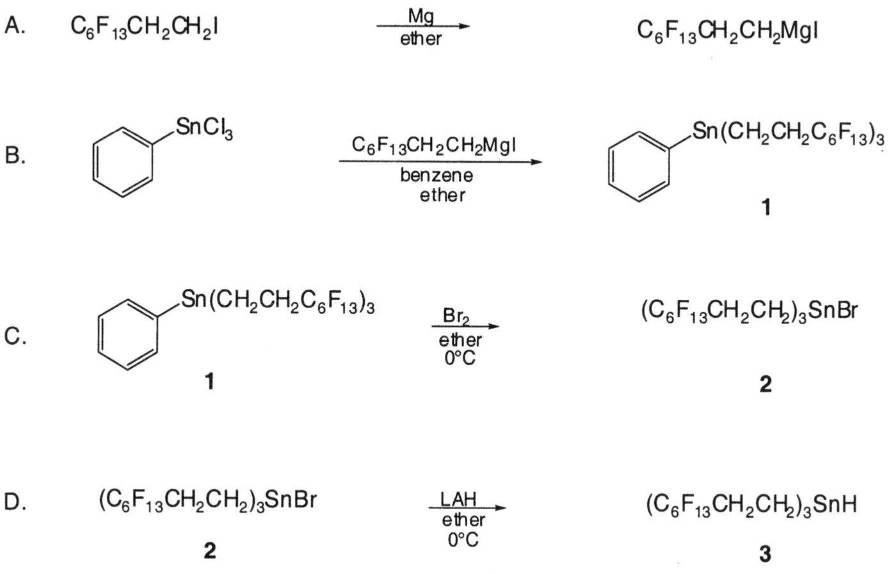

A. $C_6F_{13}CH_2CH_2I$ $\xrightarrow[\text{ether}]{\text{Mg}}$ $C_6F_{13}CH_2CH_2MgI$

B. $\xrightarrow[\substack{\text{benzene} \\ \text{ether}}]{C_6F_{13}CH_2CH_2MgI}$

C. $\xrightarrow[\substack{\text{ether} \\ 0°C}]{Br_2}$ $(C_6F_{13}CH_2CH_2)_3SnBr$

 2

D. $(C_6F_{13}CH_2CH_2)_3SnBr$ $\xrightarrow[\substack{\text{ether} \\ 0°C}]{LAH}$ $(C_6F_{13}CH_2CH_2)_3SnH$

  2   **3**

Submitted by Aimee Crombie, Sun-Young Kim, Sabine Hadida, and Dennis P. Curran.[1]

Checked by Peter Ranslow and Louis S. Hegedus.

1. Procedure

A. (Perfluorohexyl)ethylmagnesium iodide. A 500-mL, three-necked flask equipped with a stirring bar and a reflux condenser is dried in an oven overnight and then cooled under argon. Dry ether (20 mL) and 2-perfluorohexyl-1-iodoethane (1 mL) are added to magnesium (2.91 g, 120 mmol) in the dried flask equipped with a reflux condenser, thermometer and an outlet to argon gas (Note 1). The reaction is initiated by sonication for 30 min. Additional dry ether (70 mL) is added to the mixture while stirring. In a separate, dry, 100-mL, round-bottomed flask cooled under argon, dry ether (45 mL) is combined with 2-perfluorohexyl-1-iodoethane (13.70 mL, total of 60 mmol). This separate mixture is slowly added to the reaction mixture over 1 hr with stirring. The addition rate is adjusted to keep a constant temperature of about 30°C. The reaction mixture is heated at reflux for 2.5 hr in an oil bath at 50°C and allowed to stand after removal from the bath until it reaches room temperature.

B. *Tris(2-perfluorohexylethyl)phenyltin.* Phenyltin trichloride (2.46 mL, 15 mmol) is dissolved in dry benzene (30 mL) in a 100-mL, round-bottomed flask under argon at room temperature. The solution is slowly added to the 500-mL, three-necked flask containing the Grignard reagent at room temperature over 1 hr while stirring. The addition rate is adjusted to keep a constant temperature of about 25°C. The reaction mixture is heated at reflux overnight in an oil bath at 50°C, removed from the bath, and allowed to stand at ambient temperature for 4.5 hr with stirring. The reaction mixture is diluted with ether (100 mL), vacuum filtered into a 1-L Erlenmeyer flask, and hydrolyzed with saturated ammonium chloride solution (300 mL). Excess magnesium solid is also hydrolyzed with saturated ammonium chloride (100 mL) separately (Note 2). The mixture is transferred to a 1-L separatory funnel. The water layer is removed, and the organic layer is washed three times with 3% sodium thiosulfate (3 x 200 mL). The organic layer is dried over magnesium sulfate and filtered under vacuum. The solvent is evaporated to dryness under reduced pressure using a rotovap. The impure product is redissolved in ether (20 mL) and transferred to a 50-mL pear-shaped flask. The ether is removed under reduced pressure. Kugelrohr distillation is peformed to remove a dimer impurity of $(C_6F_{13}CH_2CH_2CH_2CH_2C_6F_{13})$ at 0.02 mm, 100-120°C for 5 hr (Note 3). The residue is further purified by column filtration over silica (30 g) under pressure with hexane (1 L) (Note 4). The solvent is evaporated under reduced pressure to leave 17.2 g (13.9 mmol, 93%) of pure compound as a colorless oil (Notes 5 and 9).

2

C. Bromotris[2-(perfluorohexyl)ethyl]tin. The fluorous phenyltin product (17.2 g, 13.9 mmol) and dry ether (80 mL) are transferred to a 250-mL, three-necked flask that had been dried in an oven and cooled to 0°C under argon. Bromine (0.71 mL, 14 mmol) is added dropwise over 30 min to the mixture. The addition rate is adjusted to keep the temperature between 0° and 1°C. The mixture is warmed to 25°C and stirred for 7 hr. The reaction mixture is transferred to a 250-mL, round-bottomed flask. The ether and excess bromine are removed under reduced pressure to leave a yellow oil. The oil is dissolved in FC-72 (75 mL) and transferred to a 250-mL separatory funnel. The bromine and bromobenzene by-products are removed by washing three times with methylene chloride (3 x 75 mL) leaving the fluorous layer colorless. The FC-72 is removed under reduced pressure to provide 15.8 g (12.7 mmol, 92%) of a colorless oil (Note 6).

D. Tris[(2-perfluorohexyl)ethyl]tin hydride (Note 7). A 1-L, three-necked flask and a stirring bar are dried in an oven. The fluorous tin bromide (13.8 g, 11.1 mmol) is dissolved in dry ether (275 mL) and transferred to the dried three-necked flask equipped with a thermometer, stirring bar, and an outlet to argon. The solution is cooled to 0°C. A 1 M solution of lithium aluminum hydride (LAH) in ether (11.1 mL, 11.1 mmol) is added dropwise over 45 min to the solution. The addition rate is adjusted to maintain a temperature between 0° and 1°C. The reaction mixture is stirred for 6 hr at 0°C. Water (75 mL) is slowly added (initially dropwise) with stirring to the ice-cold mixture. Sodium potassium tartrate (20%) (250 mL) is added and the mixture is transferred to a 1-L separatory funnel. The ethereal layer is separated and the aqueous layer is extracted three times with ether (3 x 100 mL). The combined extracts are dried with magnesium sulfate and vacuum filtered into a 1-L, round-bottomed flask. The solvent is evaporated under reduced pressure. The crude product is distilled under a reduced pressure of 0.02 mm at 133-140°C to provide 11.3 g (9.69 mmol, 87%) of the pure product as an oil (Notes 8 and 9).

2. Notes

1. Ether and benzene were distilled with sodium/benzophenone prior to use. The 2-perfluorohexyl-1-iodoethane was purchased from Lancaster and the FC-72 was purchased from 3M. Magnesium (powder, 50 mesh) and all other reagents were purchased from Aldrich Chemical Company, Inc.

2. The mixture can be hydrolyzed without filtration. However, it is more convenient to remove the solid magnesium and hydrolyze the two components separately.

3. A cooled collection flask and a guard collection flask were used during the Kugelrohr distillation so that the dimer impurity (white solid) would not contaminate the vacuum pump. Periodic heating of the neck of the guard flask was performed with a heat gun to prevent any blockage from the impurity. A high vacuum pump was used to reduce the pressure. Although simple distillation has been used in the past, the Kugelrohr distillation is more advantageous and more convenient.

4. Short column chromatography can be performed to purify the compound further if desired.

5. The spectral properties of product **1** are as follows: ^{1}H NMR (CDCl$_3$) δ: 1.31 [t, 6 H, J = 8.3, ^{2}J(^{119}Sn-H) = 53.4], 2.31 (m, 6 H), 7.41 (s, 5 H); ^{119}Sn NMR (CDCl$_3$) - 11.7 ppm; IR (thin film) cm^{-1}: 3100, 2950, 1238, 1190, 1144, 655; MS (m/z) 1161 (M+ - Ph), 891 (M+ - CH$_2$CH$_2$C$_6$F$_{13}$).

6. The spectral properties of product **2** are as follows: ^{1}H NMR (CDCl$_3$) δ: 1.56 [t, 6 H, J = 8.3, ^{2}J(^{119}Sn-H) = 53.4], 2.42 (m, 6 H); ^{119}Sn NMR (hexane-C$_6$D$_6$) 109.2 ppm; IR (thin film) cm^{-1}: 3600, 1250, 1227, 1145, 534; MS (m/z): 1161 (M+ - Br), 893 (M+ - CH$_2$CH$_2$C$_6$F$_{13}$).

7. Reactions on a smaller scale tended to give better yields for the reduction of the fluorous tin bromide to fluorous tin hydride.

8. The spectral properties of product **3** are as follows (Note 10): ^{1}H NMR (CDCl$_3$) δ: 1.16 [t, 6 H, J = 8.1, ^{2}J(^{119}Sn-H) = 53.4], 2.35 (m, 6 H), 5.27 (s, 1 H); ^{119}Sn NMR (CDCl$_3$) - 84.5 (^{1}J(^{119}Sn-H) = 1835); IR (thin film) cm^{-1}: 1842, 1197; MS (m/z) 1161 (M$^+$ - H), 813 (M$^+$ - CH$_2$CH$_2$C$_6$F$_{13}$).

9. Thin layer chromatography was performed using silica plates and eluting with hexane. Potassium permanganate was used to visualize the spots. The R$_f$ values for products **1** and **3** were 0.38 and 0.37, respectively.

10. All NMR samples were dissolved in chloroform. The fluorous tin hydride is only slightly soluble in chloroform. Therefore it is necessary to saturate this NMR sample. The NMR spectrum must be recorded quickly since the tin hydride reduces chloroform on standing in the light.

Waste Disposal Information

All toxic materials were disposed of in accordance with "Prudent Practices in the Laboratory"; National Academy Press; Washington, DC, 1995.

3. Discussion

Trialkyltin hydrides represent an important class of reagents in organic chemistry because of their utility in radical reactions.[2] However, problems of toxicity and the difficulty of product purification made trialkyltin hydrides less than ideal reagents.[3] Several workup procedures[4] and structurally modified trialkyltin hydrides[5] have been developed to facilitate the separation of tin residues from the reaction mixture. Tris(trimethylsilyl)silicon hydride[6a] has also been synthesized and is often used successfully in radical reactions. However, its reactivity is different from that of trialkyltin hydrides in a number of important respects. Other tin hydride surrogates are also available.[6b]

On the heels of work by Zhu[7] and Horváth and Rábai,[8a] perfluorocarbon solvents and fluorous reagents have been used increasingly in organic syntheses.[9] Fluorous compounds often partition preferentially into a fluorous phase in organic/fluorous liquid-liquid extraction, thus providing easy separation of the compounds. Tris[(2-perfluorohexyl)ethyl]tin hydride[9b-e] combines the favorable radical reaction chemistry of trialkyltin hydrides with the favorable separation features of fluorous compounds.

Tris[(2-perfluorohexyl)ethyl]tin hydride has three perfluorinated segments with ethylene spacers and it partitions primarily (> 98%) into the fluorous phase in a liquid-liquid extraction. This feature not only facilitates the purification of the product from the tin residue but also recovers toxic tin residue for further reuse. Stoichiometric reductive radical reactions with the fluorous tin hydride **3** have been previously reported and a catalytic procedure is also well established. [9b-e] The reduction of adamantyl bromide in BTF (benzotrifluoride)[10,11] using 1.2 equiv of the fluorous tin hydride and a catalytic amount of azobisisobutyronitrile (AIBN) was complete in 3 hr (Scheme 1). After the simple liquid-liquid extraction, adamantane was obtained in 90% yield in the organic layer and the fluorous tin bromide was separated from the fluorous phase. The recovered fluorous tin bromide was reduced and reused to give the same results. Phenylselenides, tertiary nitro compounds, and xanthates were also successfully reduced by the fluorous tin hydride. Standard radical additions and cyclizations can also be conducted as shown by the examples in Scheme 1. Hydrostannation reactions are also possible,[9e] and these are useful in the techniques of fluorous phase switching.[10] Carbonylations are also possible.[9p] Rate constants for the reaction of the fluorous tin hydride with primary radicals and acyl radicals have been measured; it is marginally more reactive than tributlytin hydrides.[9c,e]

6

Scheme 1

n-$C_{15}H_{31}$I

5 eq $\overset{}{\diagup}CO_2Me$

0.1 eq $(C_6F_{13}CH_2CH_2)_3SnBr$
1.2 eq $NaCNBH_3$, AIBN
BTF/tert-BuOH (1/1), reflux, 92%

n-$C_{15}H_{31}\diagup\diagup CO_2Me$

Ph Ph

0.1 eq $(C_6F_{13}CH_2CH_2)_3SnBr$

1.2 eq $NaCNBH_3$, AIBN
BTF/tert-BuOH (1/1), reflux, 75%

Ph Ph

$\diagup CO_2Bn$

1.5 eq $(C_6F_{13}CH_2CH_2)_3SnH$

AIBN, BTF, 80°C, 6 hr, 77%

$(C_6F_{13}CH_2CH_2)_3Sn\diagup\diagup CO_2Bn$

The preparation method reported here can be applied to the synthesis of a variety of related fluorous tin compounds. Seven more fluorous tin hydrides with the general formula of $[CF_3(CF_2)_n(CH_2)_m]_3SnH$ (n = 3, m = 2,3, n = 5, m = 3, and n = 9, m = 2) and $[CF_3(CH_2)_m CH_2CH_2]SnMe_2H$ (m = 5, 7, 9) were synthesized using this method and used for radical reactions.[9e] The fluorous phenyl tin compound **1** and related compounds have been successfully reacted in Stille coupling reactions[9f-k] demonstrating the easy purification feature of fluorous compounds. Fluorous tin bromide **2** is an important intermediate for the synthesis of various reagents including tin azide and allyl tin compounds.[9g,h] Fluorous silanes are made by similar routes.[9m-o]

Recently, the submitters have developed new separation procedures based on fluorous silica gel,[12] and the separation of fluorous compounds by solid phase extraction has become another option for compounds that are not easy to separate by liquid-liquid extraction.

1. Department of Chemistry, University of Pittsburgh, Pittsburgh, PA, 15260.

2. (a) Kuivila, H. G. *Acc. Chem. Res.* **1968**, *1*, 299; (b) Neumann, W. P. *Synthesis* **1987**, 665; (c) Curran, D. P. *Synthesis* **1988**, 489.

3. Pereyre, M.; Quintard, J. -P.; Rahm, A. "Tin in Organic Synthesis"; Butterworths: London, 1987.

4. (a) Berge, J. M.; Roberts, S. M. *Synthesis* **1979**, 471; (b) Curran, D. P.; Chang, C. -T. *J. Org. Chem.* **1989**, *54*, 3140; (c) Crich, D.; Sun, S. *J. Org. Chem.* **1996**, *61*, 7200.

5. (a) Gerlach, M.; Jördens, F.; Kuhn, H.; Neumann, W. P.; Peterseim, M. *J. Org. Chem.* **1991**, *56*, 5971; (b) Light, J.; Breslow, R. *Tetrahedron Lett.* **1990**, *31*, 2957; (c) Rai, R.; Collum, D. B. *Tetrahedron Lett.* **1994**, *35*, 6221; (d) Clive, D. L. J.; Yang, W. *J. Org. Chem.* **1995**, *60*, 2607.

6. (a) Chatgilialoglu, C. *Acc. Chem. Res.* **1992**, *25*, 188; (b) Walton, J. C. *Acc. Chem. Res.* **1998**, *31*, 99.

7. Zhu, D.-W. *Synthesis* **1993**, 953.

8. (a) Horváth, I. T.; Rábai, J. *Science* **1994**, *266*, 72; (b) Cornils, B. *Angew. Chem., Int. Ed. Engl.* **1997**, *36*, 2057; (c) Horváth, I. T. *Acc. Chem. Res.* **1998**, *31*, 641.

9. (a) Studer, A.; Hadida, S.; Ferritto, R.; Kim, S.-Y.; Jeger, P.; Wipf, P.; Curran, D. P. *Science* **1997**, *275*, 823; (b) Curran, D. P.; Hadida, S. *J. Am. Chem. Soc.* **1996**, *118*, 2531; (c) Horner, J. H.; Martinez, F. N.; Newcomb, M.; Hadida, S.; Curran, D. P. *Tetrahedron Lett.* **1997**, *38*, 2783; (d) Hadida, S.; Super, M. S.; Beckman, E. J.; Curran, D. P. *J. Am. Chem. Soc.* **1997**, *119*, 7406; (e) Curran, D. P.; Hadida, S.; Kim, S. -Y.; Luo, Z. *J. Am. Chem. Soc.* **1999**, *121*, 6607; (f)

Curran, D. P.; Hoshino, M. *J. Org. Chem.* **1996**, *61*, 6480; (g) Curran, D. P.

Hadida, S.; Kim, S. -Y. *Tetrahedron* **1999**, *55*, 8997; (h) Curran, D. P.; Luo, Z.

Med. Chem. Res. **1998**, *8*, 261; (i) Curran, D. P.; Luo, Z.; Degenkolb, P. *Bioorg.*

Med. Chem. Lett. **1998**, *8*, 2403; (j) Hoshino, M.; Degenkolb, P.; Curran, D. P. *J.*

Org. Chem. **1997**, *62*, 8341; (k) Larhed, M.; Hoshino, M.; Hadida, S.; Curran, D.

P.; Hallberg, A. *J. Org. Chem.* **1997**, *62*, 5583; (l) Spetseris, N.; Hadida, S.;

Curran, D. P.; Meyer, T. Y. *Organometallics* **1998**, *17*, 1458; (m) Studer, A.;

Curran, D. P. *Tetrahedron* **1997**, *53*, 6681; (n) Studer, A.; Jeger, P.; Wipf, P.;

Curran, D. P. *J. Org. Chem.* **1997**, *62*, 2917; (o) Curran, D. P.; Ferritto, R.; Hua,

Y. *Tetrahedron Lett.* **1998**, *39*, 4937; (p) Ryu, I.; Niguma, T.; Minakata, S.;

Komatsu, M.; Hadida, S.; Curran, D. P. *Tetrahedron Lett.* **1997**, *38,* 7883; (q)

Ryu, I.; Niguma, T.; Minakata, S.; Komatsu, M.; Luo, Z. Y.; Curran, D. P.

Tetrahedron Lett. **1999**, *40*, 2367.

10.   (a) Curran, D. P. In "Stimulating Concepts in Chemistry"; Stoddard, F.; Reinhoudt,
D.; Shibasaki, M., Ed.; Wiley-VCH: New York, 2000; p. 25; (b) Curran, D. P.
Angew. Chem., Int. Ed. Engl. **1998**, *37*, 1175; (c) Curran, D. P. *Chemtracts: Org.*
Chem. **1996**, *9*, 75.

11.   (a) Ogawa, A.; Curran, D. P. *J. Org. Chem.* **1997**, *62*, 450; (b) Maul, J. J.;
Ostrowski, P. J.; Ublacker, G. A.; Linclau, B.; Curran, D. P. In "Topics in Current
Chemistry: Modern Solvents in Organic Synthesis"; Knochel, P., Ed.; Springer-
Verlag: Berlin, 1999; Vol. 206; pp. 80.

12.   (a) Curran, D. P.; Hadida, S.; He, M. *J. Org. Chem.* **1997**, *62*, 6714; (b) Curran,
D. P.; Hadida, S.; Studer, A.; He, M.; Kim. S. -Y.; Luo, Z.; Larhed, M.; Hallberg,
M.; Linclau, B. In "Combinatorial Chemistry: A Practical Approach"; Fenniri, H.,
Ed.; Oxford University Press: Oxford, 2001; Vol. 2.

Appendix

Chemical Abstracts Nomenclature (Collective Index Number);
(Registry Number)

Tris[(2-perfluorohexyl)ethyl]tin hydride: Stannane, tris(3,3,4,4,5,5,6,6,7,7,8,8,8,-tridecafluorooctyl)- (13); (175354-32-2)

2-Perfluorohexyl-1-iodoethane: Octane, 1,1,1,2,2,3,3,4,4,5,5,6,6-tridecafluoro-8-iodo- (9); (2043-57-4)

Tris[(2-perfluorohexyl)ethyl]phenyl tin: Stannane, phenyltris(3,3,4,4,5,5,6,6,7,7,8,8,8-tridecafluorooctyl)- (13); (175354-30-0)

Phenyltin chloride: Aldrich: Phenyltin trichloride: Stannane, trichlorophenyl- (8,9); (1124-19-2)

Bromotris[(2-perfluorohexyl)ethyl]tin: Stannane, bromotris(3,3,4,4,5,5,6,6,7,7,8,8,8-tridecafluorooctyl)- (13); (175354-31-1)

Bromine (8,9); (7726-95-6)

Lithium aluminum hydride: Aluminate (1-), tetrahydro-, lithium; aluminate (1-), tetrahydro-, lithium, (T-4)- (9); (16853-85-3)

Magnesium (8,9); (7439-95-4)

1,2-METALLATE REARRANGEMENT:
(Z)-4-(2-PROPENYL)-3-OCTEN-1-OL
[3-Octen-1-ol, 4-(2-propenyl)-, (Z)-]

A.

B.   2 n-BuLi + CuBr · SMe$_2$ ⟶   n-Bu$_2$CuLi · LiBr

C.

Submitted by Krzysztof Jarowicki, Philip J. Kocienski, and Liu Qun.[1]

Checked by Christopher L. Franklin and Stephen F. Martin.

1. Procedure

A. *2,3-Dihydro-5-furyllithium.* A 250-mL, three-necked, round-bottomed flask (Note 1) equipped with a thermometer, Teflon-coated magnetic stirring bar (Note 2) and a nitrogen inlet is charged with a solution of 2,3-dihydrofuran (3.15 g, 3.4 mL, 45 mmol) (Note 3) in tetrahydrofuran (THF, 6.13 g, 6.9 mL, 90 mmol) (Note 4). The magnetically stirred

11

solution is cooled in a liquid nitrogen-ethyl acetate cooling bath to an internal temperature of -85°C whereupon tert-butyllithium (26.5 mL, 45 mmol, 1.7 M solution in pentane) (Note 5) is added dropwise via a syringe during 10 min (Note 6). After the addition the yellow semi-solid mixture is placed in an ice bath, stirred at 0-3°C for 30 min and diluted with 20 mL of diethyl ether (Note 4) to give a thick yellow suspension of 2,3-dihydro-5-furyllithium.[2]

B. *Lithium dibutylcuprate.* A 250-mL, three-necked, round-bottomed flask, equipped with a thermometer, Teflon-coated magnetic stirring bar and a nitrogen inlet is charged with freshly recrystallized copper bromide-dimethyl sulfide complex (10.2 g, 49.5 mmol) (Note 7) and anhydrous diethyl ether (85 mL). To this suspension, cooled to -80°C in a liquid nitrogen-ethyl acetate bath, is added dropwise via syringe during 20 min a solution of n-butyllithium (44.4 mL, 99 mmol, 2.23 M solution in hexanes) (Notes 5 and 8). When the addition is complete, the reaction mixture is placed in an ice bath and stirred at 0-3°C for 6 min to give a brown solution of lithium dibutylcuprate.

C. *(Z)-4-(2-Propenyl)-3-octen-1-ol.* The solution of lithium dibutylcuprate is cooled to -80°C and transferred by cannula (Note 9) under nitrogen pressure over 10 min to the stirred suspension of 2,3-dihydro-5-furyllithium cooled to -85°C (Note 10). After the transfer, the flask containing the cuprate is washed with 5 mL of diethyl ether. The temperature of the light brown suspension is allowed to rise slowly to 0°C over 3 hr (Note 11) and stirred at 0-3°C for 30 min. The brown solution is cooled to -80°C and a solution of allyl bromide (17.9 g, 12.8 mL, 148 mmol) (Note 3) in diethyl ether (20 mL) is added dropwise during 10 min via syringe (Note 12). The black mixture is left in the cooling bath to warm gradually to rt overnight (12 hr). The resultant dark gray suspension is treated with an aqueous saturated solution of ammonium chloride (50 mL) and aqueous ammonia (20 mL, 30% in water). The organic layer is separated and the water layer extracted with ether (2 x 75 mL). The combined organic extracts are dried with sodium sulfate, filtered, and concentrated under reduced pressure using a cold water bath (10-20°C). The crude product (light brown oil) is purified by flash chromatography on silica (150 g) (Note 13) eluting first

12

with hexanes-diethyl ether (5:1, 500 mL followed by 1:2, 700 mL). The residue obtained from evaporation of the second fraction is distilled using a Kugelrohr oven to give the product as a colorless oil (6.39 g, 84%), bp 150°C (bath)/0.01 mm) (Notes 14 and 15).

2. Notes

1. All glassware was dried for at least 8 hr at 80°C, assembled hot, flame-dried, and allowed to cool under nitrogen.

2. Because of the semi-solid nature of the reaction mixture the use of a large football stirring bar (19 x 51 mm) or a mechanical stirrer is recommended.

3. 2,3-Dihydrofuran was purchased from Aldrich Chemical Company, Inc., and freshly distilled from calcium hydride under nitrogen.

4. Tetrahydrofuran was distilled from sodium/benzophenone ketyl or potassium/benzophenone under nitrogen.

5. tert-Butyllithium was purchased from Aldrich Chemical Company, Inc., and titrated using 1,3-diphenylacetone tosylhydrazone.[3]

6. During the addition the temperature should not be allowed to rise above -68°C.

7. Commercial copper bromide or its dimethyl sulfide complex contains impurities that are deleterious to the reaction. Therefore, the copper(I) bromide-dimethyl sulfide complex is prepared according to the method of House[4] from copper(I) bromide generated by reduction of copper(II) bromide (Aldrich Chemical Company, Inc., 99%) with sodium sulfite.[5] Best results are obtained using copper(I) bromide-dimethyl sulfide complex freshly recrystallized according to the following procedure.

A 100-mL conical flask equipped with a condenser and a nitrogen inlet is charged with copper(I) bromide-dimethyl sulfide complex (15 g). Anhydrous dimethyl sulfide (50 mL) is added via syringe and the mixture heated gently until all the solid dissolves. The heating bath is removed and pentane (25 mL) is added to the warm solution. The solution

13

is cooled in an ice bath until crystallization is complete. The product is collected by filtration under nitrogen using a sintered glass funnel. The complex is washed with 7.5 mL of dimethyl sulfide-pentane solution (2:1, v/v) and dried at rt under a stream of nitrogen for 1 hr to give 11.35 g of pure copper(I) bromide-dimethyl sulfide complex as a white solid.

8. During the addition, the temperature should not be allowed to rise above -58°C.

9. Because of the viscosity of the solution the size of the cannula is important. A cannula with an internal diameter of 2 mm was made from stainless steel HPLC tubing.

10. During the addition the temperature should not be allowed to rise above -76°C.

11. It is important to increase the temperature slowly to minimize the formation of by-products. The checkers found that this was most easily performed using a dry ice-acetone bath.

12. During the addition the temperature should not be allowed to rise above -30°C, and the mixture turns black.

13. MN Kieselgel 60, 230-400 mesh, purchased from Machery-Nagel GmbH & Co. was used. The checkers used ICN 32-63 D 60 Å, purchased from ICN Pharmaceuticals, Inc.

14. The spectral properties are as follows: ^{1}H NMR (400 MHz, CDCl$_3$) δ: 0.90 (t, 3 H, J = 7.2), 1.24-1.44 (comp m, 4 H), 1.64 (br s, 1 H), 2.02 (t, 2 H, J = 7.2), 2.31 (dt, 2 H, J = 6.6, 7.1), 2.81 (d, 2 H, J = 6.5), 3.64 (t, 2 H, J = 6.5), 5.00 (ddt, 1 H, J = 1.5, 1.7, 10.0), 5.04 (dq, 1 H, J = 1.5, 17.1), 5.22 (t, 1 H, J =7.3), 5.76 (ddt, 1 H, J = 6.5, 10.2, 17.0); ^{13}C NMR (100 MHz, CDCl$_3$) δ: 14.1, 22.6, 30.4, 31.5, 35.0, 37.0, 62.7, 115.3, 121.1, 136.4, 140.8; IR (neat) cm^{-1}: 3340, 3072, 2928, 2866, 1631, 1456, 1052, 913; MS (CI+) m/z: 169.1586 (C$_{11}$H$_{20}$O + H requires 169.1592), 169, 151, 127 (base), 109; Anal. Calcd. for C$_{11}$H$_{20}$O: C, 78.51; H, 11.98%. Found C, 78.43; H, 11.83%.

14

15. Capillary gas chromatography indicated a purity of 94%: DB 225 (30 m x 0.32 mm x 0.25 μm); 100°C (2 min) temperature programmed at 8°C/min to 160°C; He carrier 2.3 mL/min.

Waste Disposal Information

All toxic materials were disposed of in accordance with "Prudent Practices in the Laboratory"; National Academy Press; Washington, DC, 1995.

3. Discussion

The preparation of (Z)-4-(2-propenyl)-3-octen-1-ol described here illustrates the use of 1,2-metallate rearrangements of metal carbenoids for the stereoselective synthesis of trisubstituted alkenes.[6] 1,2-Metallate rearrangements of boronate carbenoids are well known to proceed with inversion of stereochemistry at sp^3 hybridized carbon.[7] Examples involving cuprates,[8] manganates[9] and zincates[10] are also known. The present reaction, a rare example of a 1,2-metallate rearrangement involving inversion of stereochemistry at sp^2 hybridized carbon, was discovered during an investigation of the Cu(I)-catalyzed ring scission of 2,3-dihydro-5-furyllithium by butyllithium first described by Fujisawa and co-workers.[11] A mechanism involving the 1,2-metallate rearrangement of a higher order organocuprate intermediate has been postulated.[12]

The reaction has broad scope: 5-, 6-, and 7-membered ring metallated enol ethers participate equally well as do organocuprates derived from MeLi, PhLi, sec-BuLi, tert-BuLi, Me_3SnLi, and $PhMe_2SiLi$ among others. The reaction also works with Grignard reagents.[13] Some examples are given in the Table.

TABLE
DIENOLS VIA Cu(I)-MEDIATED 1,2-METALLATE REARRANGEMENT[a]

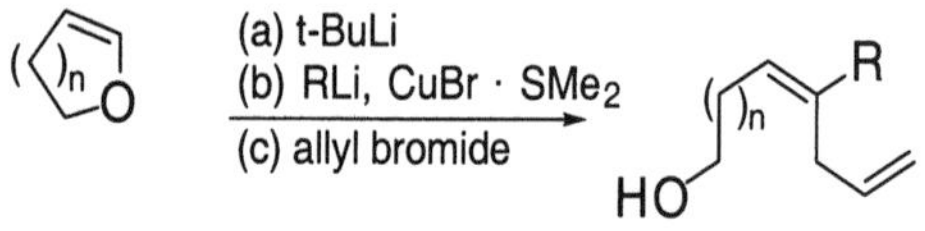

(a) t-BuLi
(b) RLi, CuBr · SMe_2
(c) allyl bromide

Entry	n	R	Time (min)[b]	Yield (%)
1	1	Me	30	81
2		Ph		81
3		sec-Bu		90
4		tert-Bu		90
5	2	n-Bu	180[c]	79
6		sec-Bu		82
7		tert-Bu		87

[a] The reactions were performed on a 5-mmol scale.
[b] Time of stirring the reaction mixture at 0-3°C after transferring the solution of the cuprate to the lithiated enol ether and warming to 0°C.
[c] Dihydropyran (n = 2) is less reactive than dihydrofuran (n = 1) and requires a longer reaction time.

The final step in the sequence can be accomplished with less reactive halides such as hexyl iodide with two modifications to the reaction conditions: the lithium dialkylcuprate must be generated in 1,2-dimethoxyethane instead of diethyl ether and HMPA must be added along with the alkylating agent as illustrated in the following example:

16

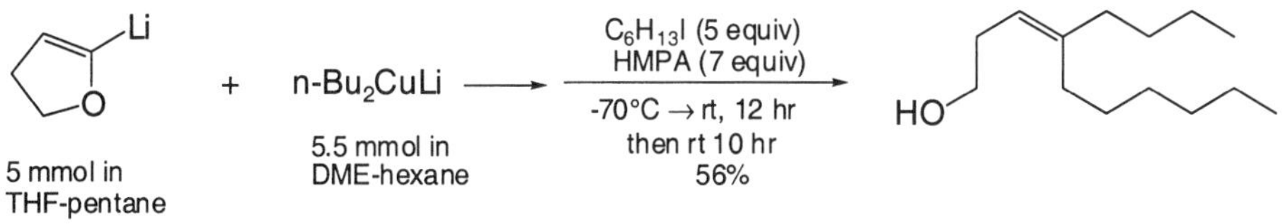

Li
+ n-Bu$_2$CuLi ⟶

C$_6$H$_{13}$I (5 equiv)
HMPA (7 equiv)
-70 °C → rt, 12 hr
then rt 10 hr
56%

5 mmol in
THF-pentane

5.5 mmol in
DME-hexane

HO

1. Department of Chemistry, University of Leeds, Leeds LS2 9JT, UK

2. Boeckman, R. K., Jr.; Bruza, K. J. *Tetrahedron* **1981**, *37*, 3997.

3. Lipton, M. F.; Sorensen, C. M.; Sadler, A. C.; Shapiro, R. H. *J. Organometal. Chem.* **1980**, *186*, 155.

4. House, H. O.; Chu, C.-Y.; Wilkins, J. M.; Umen, M. J. *J. Org. Chem.* **1975**, *40*, 1460.

5. Theis, A. B.; Townsend, C. A. *Synth. Commun.* **1981**, *11*, 157.

6. Kocienski, P.; Wadman, S.; Cooper, K. *J. Am. Chem. Soc.* **1989**, *111*, 2363.

7. Matteson, D. S. *Chem. Rev.* **1989**, *89*, 1535.

8. Yamamoto, H.; Kitatani, K.; Hiyama, T.; Nozaki, H. *J. Am. Chem. Soc.* **1977**, *99*, 5816.

9. Kakiya, H.; Inoue, R.; Shinokubo, H.; Oshima, K. *Tetrahedron Lett.* **1997**, *38*, 3275.

10. Harada, T.; Katsuhira, T.; Hattori, K.; Oku, A. *J. Org. Chem.* **1993**, *58*, 2958.

11. Fujisawa, T.; Kurita, Y.; Kawashima, M.; Sato, T. *Chemistry Lett.* **1982**, 1641.

12. Kocienski, P. In "Organic Synthesis via Organometallics"; Enders, D.; Gais, H.-J.; W. Keim, W., Eds.; Verlag Vieweg: Wiesbaden, **1993**; pp 203.

13. Barber, C.; Bury, P.; Kocienski, P.; O'Shea, M. *J. Chem. Soc., Chem. Commun.* **1991**, 1595.

Appendix

Chemical Abstracts Nomenclature (Collective Index Number); (Registry Number)

(Z)-4-(2-Propenyl)-3-octen-1-ol: 3-Octen-1-ol, 4-(2-propenyl)-, (Z)- (12); (119528-99-3)

2,3-Dihydro-5-furyllithium: Lithium, (4,5-dihydro-2-furanyl)- (9); (75213-94-4)

2,3-Dihydrofuran: Furan, 2,3-dihydro- (8,9); (1191-99-7)

tert-Butyllithium: Lithium, tert-butyl- (8); Lithium, (1,1-dimethylethyl)- (9); (594-19-4)

Copper (I) bromide-dimethyl sulfide complex: Copper, bromo[thiobis[methane]]- (9); (54678-23-8)

Allyl bromide: 1-Propene, 3-bromo- (8,9); (106-95-6)

Dimethyl sulfide: Methyl sulfide (8); Methane, thiobis- (9); (75-18-3)

Butyllithium: Lithium, butyl- (8,9); (109-72-8)

DIMETHYLTITANOCENE

[Titanium, $bis(\eta^5$-2,4-cyclopentadien-1-yl)dimethyl-]

Submitted by Joseph F. Payack,[1] David L. Hughes,[1] Dongwei Cai,[1]
Ian F. Cottrell,[2] and Thomas R. Verhoeven.[1]
Checked by Mark Kerr and Louis Hegedus.

1. Procedure[3]

A 1-L, three-necked, round-bottomed flask, equipped for mechanical stirring, and outfitted with a 250-mL, pressure equalizing addition funnel, and a Claisen adapter bearing a thermometer and a nitrogen inlet/outlet vented through a mineral oil bubbler, is placed under a nitrogen atmosphere and charged with 41.5 g (0.167 mol) of titanocene dichloride (Note 1), and 450 mL of dry toluene (Note 2). The slurry is efficiently stirred and chilled to an internal temperature of -5°C in a ice/methanol bath; then 126 mL of a 3 M solution (0.38 mol) of methylmagnesium chloride in tetrahydrofuran (THF) (Note 1) is added dropwise via the addition funnel over 1 hr, at a rate of addition adjusted to maintain an internal temperature below +8°C. The resulting orange slurry (Note 3) is mechanically stirred at an internal temperature of 0 to +5°C for 1 hr, or until the insoluble purple titanocene dichloride is no longer seen in the suspension (Note 4). The addition funnel is removed and replaced by a rubber septum, and the reaction is assayed by ^{1}H NMR (Notes 5, 6).

While the reaction is aging at 0° to +5°C, a 2-L, three-necked, round-bottomed flask, equipped for mechanical stirring, and outfitted with a rubber septum, and a Claisen adapter bearing a thermometer and a nitrogen inlet/outlet vented through a mineral oil bubbler, is placed under a nitrogen atmosphere and charged with 117 mL of 6% aqueous ammonium chloride (7.0 g diluted to 117 mL) (Note 1). The solution is chilled to 1 to 2°C, with efficient mechanical stirring. When the formation of dimethyltitanocene is judged to be complete, the toluene/THF reaction mixture is quenched into (Note 7) the well-stirred aqueous ammonium chloride solution via a cannula (Note 8) over a period of 1 hr, maintaining an internal temperature of 0° to +5°C in both flasks. Toluene (30 mL) is used to rinse the reaction flask. The biphasic mixture is then poured into a 2-L separatory funnel, with another 30 mL of toluene rinse, and the aqueous phase is separated (Note 9). The organic layer is washed sequentially with three portions of cold water (100 mL each) and brine (100 mL), then dried over anhydrous sodium sulfate (Na_2SO_4, 35 g). The organic layer is filtered and carefully (Note 10) concentrated under reduced pressure on a rotary evaporator at a bath temperature of no more than 35°C to a weight of 150 g. The resulting orange solution is assayed by ^{1}H NMR (Note 11) to be 20 weight percent dimethyltitanocene (29.55 g, 85.0%). If the solution is to be stored for more than a week, the reagent should be diluted with 160 mL of dry THF (Note 2), which has a stabilizing effect on the labile reagent (Note 12). The solution is stored at 0° to 10°C under nitrogen in a rubber septum-sealed, round-bottomed flask. It proved effective in the conversion of a complex ester to an enol ether (Note 13).

2. Notes

1. This reagent was purchased from Aldrich Chemical Company, Inc., and used without further purification.

2. This solvent was purchased from Fisher Scientific Company and dried over 3Å sieves to a water content of less than 150 μg/mL by Karl Fisher titration.

3. The sparingly soluble titanocene dichloride reacts in stepwise fashion with methylmagnesium chloride to give the soluble red $Cp_2Ti(Me)Cl$, then the soluble orange dimethyltitanocene, precipitating magnesium chloride ($MgCl_2$). The mixture becomes thick with $MgCl_2$ as the reaction proceeds and vigorous stirring is required.

4. Since the titanocene dichloride is insoluble and the intermediate $Cp_2Ti(Me)Cl$ is soluble, the second methyl group adds much faster than the first. Reaction progress can be monitored by visually observing the disappearance of the purple crystalline titanocene dichloride.

5. A ca. 200 μL-sample is removed via *wide* bore syringe or pipet (so that a representative sample is pulled from the heterogeneous reaction mixture) and is quenched into 1 mL of water. The mixture is extracted with 1 mL $CDCl_3$, the organic phase is dried with sodium sulfate, and the solution is filtered into an NMR tube. The progress is evaluated by observing the Cp singlets: the reaction is considered complete when less than 3% combined of titanocene dichloride and $Cp_2Ti(Me)Cl$ remains by ^{1}H-NMR.

6. The spectra are as follows: ^{1}H NMR ($CDCl_3$) Cp_2TiMe_2: δ 6.05 (s, 10 H), -0.05 (s, 6 H). $Cp_2TiClMe$: δ 6.22 (s, 10 H), 0.80 (s, 3 H). Cp_2TiCl_2: δ 6.56 (s, 10 H). ^{13}C NMR Cp_2TiMe_2: δ 113.20 (Cp_2), 45.77 (Me_2). $Cp_2TiClMe$: δ 115.86 (Cp_2), 50.37 (Me). Cp_2TiCl_2: δ 120.18.

7. The reaction *must* be quenched *into* 6% aq ammonium chloride, or substantial decomposition will occur. The amount of ammonium chloride is optimized.

8. The reaction mixture is very thick and is too viscous to flow through a standard 12-gauge cannula. The transfer is best accomplished by using 1/4" to 3/8" i.d. inert tubing (PTFE, polytetrafluoroethylene, or polypropylene) inserted through both

21

septa. The mixture can then be transferred by applying only slight positive pressure on the reaction flask.

9. The workup may be done in a normal separatory funnel under air.

10. The solution must *not* be allowed to evaporate to dryness: dimethyltitanocene is unstable in the solid phase and could decompose with heat and gas evolution. The compound is also known to be unstable in neat solution at temperatures above 60°C. The concentration must be done at high (20 mm or less) vacuum.

11. The ^{1}H NMR weight percent assay is done by diluting several drops of the solution into 1 mL of $CDCl_3$ and running the spectrum with a 10-sec relaxation delay between pulses to ensure an accurate integration. The product Cp singlet is integrated against the toluene methyl singlet and any residual THF.

12. The submitters have some evidence that thermally stressed solutions of dimethyltitanocene are more stable when diluted with an equal volume of THF. No complete study of the long term 5°C stability of the solution has been done, but a 10 wt% solution in THF/toluene can be stored for several months in the refrigerator.

13. A 50-mL, nitrogen-purged round-bottomed flask was charged with cis-ester **1** ((2R-cis)-3-(4-fluorophenyl)-4-benzyl-2-morpholinyl 3,5-bis(trifluoromethyl)benzoate) (2.41 g, 4.57 mmol), dimethyltitanocene in toluene (12 mL of a 20% w/w solution in toluene), and titanocene dichloride (71 mg, 0.28 mmol). The red/orange mixture was heated to 80°C and was aged in the dark for 5.5 hr, then cooled to ambient temperature. Sodium bicarbonate (0.60 g), methanol (9.6 mL) and water (0.36 mL) were added, and the mixture was heated to 40°C for 14hr. (The hot aqueous methanol treatment was done to decompose the titanium residues into an insoluble solid. The decomposition was judged to be complete when gas evolution ceased.) The green mixture was cooled to ambient temperature and the titanium residues were removed by filtration. The solution was evaporated under reduced pressure and flushed with methanol. The crude material was recrystallized by dissolving in hot (60°C) methanol

(24 mL), cooling to ambient temperature, then adding water (7.2 mL) over 2 hr. The material was stirred for 18 hr then isolated via filtration at ambient temperature. The filtercake was washed with 25% aq methanol (6 mL) and the solid was dried at ambient temperature under nitrogen. Vinyl ether **2** ((2R-cis)-2-[[1-[3,5-bis(trifluoromethyl)phenyl]ethenyl]oxy]-3-(4-fluorophenyl)-4-benzylmorpholine) (2.31 g, 96%) was isolated as a pale yellow solid.

Waste Disposal Information

All toxic materials were disposed of in accordance with "Prudent Practices in the Laboratory"; National Academic Press; Washington, DC, 1996.

3. Discussion

Petasis, et al. have discovered that dimethyltitanocene is an excellent substitute for the Tebbe reagent[4] for the methylenation of heteroatom-substituted carbonyl compounds.[5,6]

$$\underset{\substack{\text{'R}}}{\overset{\text{R}-\text{X}}{\diagdown}}\text{C}=\text{O} \xrightarrow[\text{X = C, O, N}]{\text{Cp}_2\text{TiMe}_2, \Delta} \underset{\substack{\text{'R}}}{\overset{\text{R}-\text{X}}{\diagdown}}\text{C}=\text{CH}_2$$

The advantages of this reagent are its straightforward synthesis and relative air stability. The previous procedure[7] for the synthesis of dimethyltitanocene used methyllithium in diethyl ether, which is unsuitable for large scale operations because of its extreme pyrophoricity.[8] In addition, the method isolated the compound as a crystalline solid, which the submitters have found to be very unstable.[9] The method described here addresses both of these concerns, and can be used to prepare multiple kilograms of the reagent.

1. Department of Process Research, Merck Research Laboratories, Division of Merck & Co., Inc. P. O. Box 2000, Rahway, NJ 07065.

2. Development Laboratories Merck Research Laboratories, Division of Merck & Co., Inc. Hertford Road, Hoddesdon, Hertfordshire, EN11 9BU, England.

3. A less detailed version of this procedure has been published: Payack, J. F.; Hughes, D. L.; Cai, D.; Cottrell, I. F.; Verhoeven, T. R. *Org. Prep. Proced., Int.* **1995**, *27*, 707.

4. (a) Pine, S. H.; Zahler, R.; Evans, D. A.; Grubbs, R. H. *J. Am. Chem. Soc.* **1980**, *102*, 3270; (b) Pine, S. H.; Kim, G.; Lee, V. *Org. Synth.* **1990**, *69*, 72.

5. (a) Petasis, N. A.; Bzowej, E. I. *J. Am. Chem. Soc.* **1990**, *112*, 6392; (b) Petasis, N. A.; Lu, S.-P. *Tetrahedron Lett.* **1995**, *36*, 2393.

6. For some applications see: (a) Petasis, N. A.; Patane, M. A. *Tetrahedron Lett.* **1990**, *31*, 6799; (b) DeShong, P.; Rybczynski, P. J. *J. Org. Chem.* **1991**, *56*, 3207; (c) Swenton, J. S.; Bradin, D.; Gates, B. D. *J. Org. Chem.* **1991**, *56*, 6156; (d) Petasis, N. A.; Bzowej, E. I. *Tetrahedron Lett.* **1993**, *34*, 1721; (e) Chenault, H. K.; Chafin, L. F. *J. Org. Chem.* **1994**, *59*, 6167; (f) Kuzmich, D.;

Wu, S. C.; Ha, D. -C.; Lee, C. -S.; Ramesh, S.; Atarashi, S.; Choi, J. -K.; Hart, D. J. *J. Am. Chem. Soc.* **1994**, *116*, 6943.

7. Clauss, K.; Bestian, H. *Justus Liebigs Ann. Chem.* **1962**, *654*, 8.

8. The preparation of dimethyltitanocene from methylmagnesium iodide and dichlorotitanocene in diethyl ether has been reported in a German patent. Few details of the procedure are provided, and a yield of 58% is reported. German Patent #1,037,446 (March 12, 1959) to Farbwerke Hoechst (*Chem. Abstr.*, **1960**, *54*:18546f;).

9. For a discussion of the solid state stability of dimethyltitanocene, see: Erskine, G. J.; Hartgerink, J.; Weinberg E. L.; McCowan, J. D. *J. Organomet. Chem.* **1979**, *170*, 51 and references therein.

Appendix
Chemical Abstracts Nomenclature (Collective Index Number);
(Registry Number)

Dimethyltitanocene: Titanium, di-π-cyclopentadienyldimethyl- (8); Titanium, bis(η^5-2,4-cyclopentadien-1-yl)dimethyl- (9); (1271-66-5)

Titanocene dichloride: ALDRICH: Bis(cyclopentadienyl)titanium dichloride: Titanium, dichloro-π-cyclopentadienyl- (8); Titanium, dichlorobis(η^5-2,4-cyclopentadienyl-1-yl)- (9); (1271-19-8)

Methylmagnesium chloride: Magnesium, chloromethyl- (8, 9); (676-58-4)

Chloromethyltitanocene: Titanium, chlorodi-π-cyclopentadienylmethyl- (8); Titanium, chlorobis(η^5-2,4-cyclopentadien-1-yl)methyl- (9); (1278-83-7)

Benzoic acid, 3,5-bis(trifluoromethyl)-, (2R,3S)-3-(4-fluorophenyl)-4-(phenylmethyl)-2-morpholinyl ester (9); (170729-77-8)

Morpholine, 2-[[1-[3,5-bis(trifluoromethyl)phenyl]ethenyl]oxy]-3-(4-fluorophenyl)-4-(phenylmethyl)-, (2R,3S)- (9); (170729-78-9)

MOLYBDENUM CARBONYL-CATALYZED ALKYNOL CYCLOISOMERIZATION: PREPARATION OF 2-PHENYL-2,3-DIHYDROFURAN

(Furan, 2,3-dihydro-2-phenyl-)

Submitted by Frank E. McDonald and Brian H. White.[1]

Checked by Peter B. Ranslow and Louis S. Hegedus.

1. Procedure

Caution: All manipulations should be conducted in a well-ventilated fume hood.

A. *1-Phenyl-3-butyn-1-ol* (**1**) (Note 1). A 1000-mL, oven-dried, three-necked, round-bottomed flask is equipped with a magnetic stir bar and pressure-equalizing addition funnel, fitted with a rubber septum, and placed under an argon atmosphere. The flask is charged with lithium acetylide-ethylenediamine complex (50 g, 543 mmol) (Note 2), which is dissolved in anhydrous dimethyl sulfoxide (360 mL) (Note 3) with stirring. The flask is placed in a room temperature water bath (Note 4), the addition funnel is charged with styrene oxide (42.0 mL, 368 mmol) (Note 5), and styrene oxide is added dropwise over a period of approximately 5 min. The reaction mixture is stirred for 2 hr and quenched b y

27

pouring slowly into 600 mL of ice water in a 4-L beaker (Note 6). The contents are transferred to a 2-L separatory funnel, and the mixture is extracted with diethyl ether (6 x 350 mL). The combined organic extracts are washed once with water and decanted with evaporation of solvents by rotary evaporation. The crude product is purified by vacuum distillation. Any remaining traces of solvent and water distill over first, followed by the product (88-89°C, 1.0 mm) to provide 1-phenyl-3-butyn-1-ol (**1**, 43.98 g, 82% yield) as a colorless oil.

   B. 2-Phenyl-2,3-dihydrofuran (**2**). A 500-mL, oven-dried Airfree® reaction flask (Note 7) containing a magnetic stir bar is charged with molybdenum hexacarbonyl ($Mo(CO)_6$, 3.21 g, 12.2 mmol) (Note 8) and fitted with a rubber septum, with an argon atmosphere introduced via the side-arm. Triethylamine (Et_3N, 220 mL, 1.58 mol) (Note 9) is added, followed by diethyl ether (Et_2O, 180 mL) (Note 10), and the mixture is stirred for 10-15 min, until the molybdenum hexacarbonyl has dissolved. The solution is placed in a Rayonet Photochemical Reactor Chamber (Notes 11, 12, 13) equipped with 350-nm ultraviolet lamps. The septum is removed under positive argon pressure, and a reflux condenser bearing a rubber septum that has been previously flushed with argon is quickly fitted onto the Schlenk tube. The solution is irradiated for 1 hr under argon while the photochemical reactor interior is cooled with the built-in cooling fan. The light is turned off and the reaction mixture is allowed to cool to room temperature while maintaining an inert atmosphere to afford a yellow solution of triethylamine-molybdenum pentacarbonyl. The condenser is removed and a septum is quickly refitted while under a positive flow of argon. A solution of 1-phenyl-3-butyn-1-ol (**1**, 13.50 g, 92.3 mmol) in diethyl ether (40 mL) is injected into the solution, and the mixture is stirred at room temperature under a slow argon stream for 72 hr; the solution slowly turns dark red over this period. The solvent is then removed by rotary evaporation, leaving a dark red liquid and a precipitate of molybdenum-containing by-products, which are removed by sublimation by heating under vacuum (35°C, 0.5 mm) The remaining liquid is vacuum distilled through a short-path distillation column (45-47°C, 0.5 mm) to give 2-phenyl-2,3-dihydrofuran (**2**, 10.3 g, 76% yield) as a clear liquid (Notes 14,15).

2. Notes

1. This preparation was previously described by Brandsma.[2] Substrate **1** can be prepared in enantiomerically pure form beginning with chiral, non-racemic styrene oxide, available as either antipode from Aldrich Chemical Company, Inc., or by kinetic resolution of racemic styrene oxide.[3]

2. Lithium acetylide-ethylenediamine complex was purchased from the Aldrich Chemical Company, Inc., and used as received.

3. Anhydrous dimethyl sulfoxide was purchased from the Aldrich Chemical Company, Inc., and used as received in a Sure/Seal bottle.

4. This reaction is somewhat exothermic, and the water bath serves as a heat sink to maintain the reaction temperature under 25°C.

5. Styrene oxide was purchased from the Aldrich Chemical Company, Inc., and used without purification. *Caution! Styrene oxide is listed as a cancer suspect agent.*

6. The quench should be done by pouring the reaction mixture *very slowly* into ice water, as the quench is rather violent on occasions when unreacted lithium acetylide is present.

7. This glass flask is Kjeldahl-shaped, with a ground glass 24/40 top joint and side-arm fitted with a 2-mm ground glass stopcock, and was purchased from Chemglass (part number AF-0520-08), 3861 North Mill Road, Vineland, NJ 08360, phone 1-800-843-1794.

8. Molybdenum hexacarbonyl was purchased from Aldrich Chemical Company, Inc., and used without further purification.

9. Triethylamine was purchased from Fisher Scientific Company, and purified immediately before use by distillation from calcium hydride under an inert atmosphere.

10. Diethyl ether was purchased from Mallinckrodt Baker, Inc., and purified immediately before use by distillation from sodium/benzophenone under inert atmosphere.

11. The model used was a RPR-100 reactor purchased from the Southern New England Ultraviolet Company, Branford, CT.

12. The photochemical step could also be accomplished by adding $Mo(CO)_6$, Et_3N and Et_2O to a photochemical immersion well, and irradiating under an inert atmosphere with a Hanovia medium pressure 450W mercury vapor lamp for 20-30 min. Commercially available immersion wells generally require > 800 mL of solvent in order to work effectively (so that the solvent is level with the lamp), and the submitters have found that the more concentrated solution reported here (ca. 450 mL) is more effective.

13. If photochemical apparatus is not available, the cycloisomerization reaction can be conducted using trimethylamine N-oxide to promote oxidative decarbonylation of molybdenum hexacarbonyl in a mixture of Et_3N and Et_2O, followed by addition of 1-phenyl-3-butyn-1-ol (1).[4a] In the submitters' hands, this procedure required somewhat higher loading of molybdenum hexacarbonyl, and purification of the 2-phenyl-2,3-dihydrofuran (2) product required silica gel chromatography.

14. 2-Phenyl-2,3-dihydrofuran (2) tends to turn pale yellow after cooling and exposure to air, but no significant decomposition is revealed by NMR.

15. Characterization data for compound 2: IR (neat) cm^{-1}: 3053, 2923, 2858, 1620, 1493, 1451, 1136, 1051, 930, 782, 697; ^1H NMR (400 MHz, $CDCl_3$) δ: 2.59-2.66 (m, 1 H), 3.06-3.13 (m, 1 H), 4.97 (q, 1 H, J = 2.8), 5.53 (dd, 1 H, J = 2, 8.4), 6.47 (q, 1 H, J = 2), 7.28-7.38 (m, 5 H); ^{13}C NMR (100 MHz, $CDCl_3$) δ: 38.1, 82.5, 99.3, 125.8, 127.8, 128.7, 143.2, 145.5; MS (70 eV, EI) 146, 117, 105, 91, 77, 57, 43; HRMS (EI) calcd for $C_{10}H_{10}O$ 146.0732; found 146.0742. Anal. Calcd for $C_{10}H_{10}O$: C, 82.16; H, 6.90. Found: C, 82.07; H, 6.88.

Waste Disposal Information

All toxic materials were disposed of in accordance with "Prudent Practices in the Laboratory"; National Academy Press; Washington, DC, 1995.

3. Discussion

The single-step transformation of alkynyl alcohols to endocyclic enol ethers was unknown until the submitters' discovery that trialkylamine-molybdenum pentacarbonyl reagents catalyzed the cycloisomerization of 1-phenyl-3-butyn-1-ol (1) into 2-phenyl-2,3-dihydrofuran (2).[4] Cycloisomerization of alkynol 1 to dihydrofuran 2 was previously accomplished by multistep synthesis, including hydroboration/oxidation of the alkyne of 1 followed by hemiacetal acylation and thermal elimination.[5] A two-step preparation of 2 involving the stoichiometric reaction of 1 with chromium pentacarbonyl-diethyl ether complex and subsequent thermal reaction with dimethylaminopyridine was reported after their initial communications.[6] The title compound 2 has also been prepared by pyrolysis of 1-phenyl-2-vinyloxirane at 450°C and 15 mm.[7] Palladium-catalyzed Heck reactions of 2,3-dihydrofuran with iodobenzene or phenyl triflate provide compound 2 along with the 2,5-dihydro regioisomer, although regioselectivity can be enhanced for either isomer depending on the choice of ligands and additives.[8] An enantioselective Heck synthesis of title compound 2 has also been reported.[9]

The molybdenum-catalyzed cyclization procedure works well for a variety of homoprogargylic alcohols to afford the cycloisomeric 2,3-dihydrofuran compounds, as shown in Table I. The transformation was originally discovered with the reagent arising from reaction of molydbenum hexacarbonyl and trimethylamine oxide,[4a] but catalyst turnover and product isolation yields are significantly improved with the current procedure, which involves photolysis of molybdenum hexacarbonyl in the presence of excess triethylamine prior to addition of the alkynyl alcohol substrate.[4b] Chiral non-racemic alkynyl alcohol substrates undergo cycloisomerization without racemization at stereogenic centers (entries 2, 4-7).[10,11] The method is compatible with ester, amide, and silyl ether functional groups, and five-membered ring products are generally produced in good yields. The submitters have observed that good leaving groups at the propargylic position tend to provide furan products by a cyclization/elimination process (entries 11-13).[4,11]

The molybdenum-catalyzed alkynol cycloisomerization is the key transformation in short, stereoselective syntheses of the anti-AIDS drug d4T,[10] the antibiotic cordycepin,[10] and puromycin aminonucleoside.[11] Reaction in the presence of tributyltin triflate affords the corresponding 5-tributylstannyl-2,3-dihydrofuran products (entries 8-10).[12] Tungsten carbonyl-catalysis has recently been demonstrated for the efficient cycloisomerization of bishomopropargylic alcohols to the corresponding six-membered ring dihydropyran products (entry 7).[13] Analogous cycloisomerization reactions of terminal alkynes tethered to nitrogen,[14] carbon,[15] and sulfur[16] nucleophiles have also been developed.

1. Department of Chemistry, Emory University, 1515 Pierce Drive, Atlanta, GA 30322.

2. Brandsma, L. "Preparative Acetylenic Chemistry", 2nd ed.; Elsevier: Amsterdam, 1988; p. 67.

3. Tokunaga, M.; Larrow, J. F.; Kakiuchi, F.; Jacobsen, E. N. *Science* **1997**, *277*, 936.

4. (a) McDonald, F. E.; Connolly, C. B.; Gleason, M. M.; Towne, T. B.; Treiber, K. D. *J. Org. Chem.* **1993**, *58*, 6952; (b) McDonald, F. E.; Schultz, C. C. *J. Am. Chem. Soc.* **1994**, *116*, 9363.

5. Dana, G.; Figadére, B.; Touboul, E. *Tetrahedron Lett.* **1985**, *26*, 5683.

6. Schmidt, B.; Kocienski, P.; Reid, G. *Tetrahedron* **1996**, *52*, 1617.

7. Paladini, J. C.; Chuche, J. *Tetrahedron Lett.* **1971**, 4383.

8. (a) Larock, R. C.; Gong, W. H. *J. Org. Chem.* **1990**, *55*, 407; (b) Jeffery, T.; David, M. *Tetrahedron Lett.* **1998**, *39*, 5751.

9. Ozawa, F.; Kubo, A.; Hayashi, T. *J. Am. Chem. Soc.* **1991**, *113*, 1417.

10. McDonald, F. E.; Gleason, M. M. *Angew. Chem. Int. Ed. Engl.***1995**, *34*, 350.

11. McDonald, F. E.; Gleason, M. M. *J. Am. Chem. Soc.* **1996**, *118*, 6648.

12. McDonald, F. E. Schultz, C. C.; Chatterjee, A. K. *Organometallics* **1995**, *14*, 3628.

13. McDonald, F. E.; Reddy, K. S.; Díaz, Y. *J. Am. Chem. Soc.* **2000**, *122*, 4304.

14. McDonald, F. E.; Chatterjee, A. K. *Tetrahedron Lett.* **1997**, *38*, 7687.

15. McDonald, F. E.; Olson, T. C. *Tetrahedron Lett.* **1997**, *38*, 7691.

16. McDonald, F. E.; Burova, S. A.; Huffman, L. G., Jr. *Synthesis* **2000**, 970.

17. McDonald, F. E.; Zhu, H. Y. H. *Tetrahedron* **1997**, *53*, 11061.

Appendix

Chemical Abstracts Nomenclature (Collective Index Number); (Registry Number)

2-Phenyl-2,3-dihydrofuran: Furan, 2,3-dihydro-2-phenyl- (8,9); (33732-62-6)

1-Phenyl-3-butyn-1-ol : Benzenemethanol, α-2-propynyl- (9); (1743-36-8)

Lithium acetylide-ethylenediamine complex: Ethylenediamine, compd. with lithium acetylide (Li(HC$_2$)) (1:1) (8); 1,2-Ethanediamine, compd. with lithium acetylide (Li(HC$_2$)) (1:1) (9); (6867-30-7)

Dimethyl sulfoxide: Methyl sulfoxide (8); Methane, sulfinylbis- (9); (67-68-5)

Styrene oxide: Benzene, (epoxyethyl)- (8); Oxirane, phenyl- (9); (96-09-3)

Molybdenum hexacarbonyl: Molybdenum carbonyl (8); Molybdenum carbonyl, (OC-6-11) (9); (13939-06-5)

Triethylamine (8); Ethanamine, N,N-diethyl- (9); (121-44-8)

TABLE I
PREPARATION OF ENDOCYCLIC ENOL ETHERS VIA ALKYNOL CYCLOISOMERIZATION

Entry	Alkynyl alcohol	Endocyclic enol ether	Isolated yield	Reference
1			89%[a] (4.6 mmol)	4b
			80%[a] (75 mmol)	(this work)
			71%[b]	4a
2			80%[b]	10, 11
3			53%[b]	4a
4			89%[a]	11
5			92%[a]	11
6			35%[a] (+ 49% recovered alkynol)	17
7			98%[c]	13
8			64%[d]	12
9			65%[d]	12
10			45%[d]	12
11			60%[a]	11
12			76%[a]	11
13			85%[a]	4b

[a] cat. $(Et_3N)Mo(CO)_5$, Et_3N, Et_2O. [b] cat. $Mo(CO)_6$, Me_3NO, Et_3N, Et_2O.
[c] cat. $W(CO)_6$, Et_3N, Et_2O, hv (350 nm), [d] cat. $(Et_3N)Mo(CO)_5$, Bu_3SnOTf, Et_3N, Et_2O

ETHYL 3-(p-CYANOPHENYL)PROPIONATE FROM ETHYL 3-IODOPROPIONATE AND p-CYANOPHENYLZINC BROMIDE

[Benzenepropanoic acid, 4-cyano-, ethyl ester]

A.

$$\text{EtO–CO–CH}_2\text{CH}_2\text{Cl} \quad \xrightarrow[\text{80°C, 16 hr}]{\text{NaI, acetone}} \quad \text{EtO–CO–CH}_2\text{CH}_2\text{I}$$

B.

$$\text{Br–C}_6\text{H}_4\text{–CN} \quad \xrightarrow[\text{2) ZnBr}_2\text{, -78°C}]{\text{1) n-BuLi, -100°C, 30 min}} \quad \text{BrZn–C}_6\text{H}_4\text{–CN}$$

C.

ethyl 3-iodopropionate
p-cyanophenylzinc bromide, 2.5 equiv.
Ni(acac)$_2$ 10 mol %
4-fluorostyrene 20 mol %
THF/NMP 2:1

Submitted by Anne Eeg Jensen, Florian Kneisel, and Paul Knochel.[1]

Checked by V. Girijavallabhan and Marvin J. Miller

1. Procedure

A. Ethyl 3-iodopropionate. A 1-L, round-bottomed flask equipped with a magnetic stirring bar and a reflux condenser is charged with ethyl 3-chloropropionate (27.3 g, 0.2 mol) (Note 1) and acetone (400 mL). Sodium iodide (300 g, 2 mol) (Note 2) is added to the clear solution and the mixture is refluxed for 16 hr. The resulting pale yellow reaction mixture is cooled to room temperature, the stirring bar and reflux condenser are removed and the acetone is removed on a rotary evaporator at 40°C/550 mbar (412 mm). The residue is taken up in diethyl ether (300 mL) and washed with a saturated aqueous solution of sodium thiosulfate (3 x 100 mL). The ethereal phase is dried over anhydrous magnesium sulfate, filtered and the ether is removed by rotary evaporation at 40°C. The resulting yellow oil is purified by distillation with a membrane pump 90°C/25 mbar (18.7 mm) yielding 36.8 g of ethyl 3-iodopropionate as a clear oil (82%) (Note 3).

B. 4-Cyanophenylzinc bromide. A dry, 250-mL, three-necked flask equipped with an argon inlet and a stirring bar is charged with 4-bromobenzonitrile (9.1 g, 50 mmol) (Note 4) and evacuated for 5 min. The flask is flushed with argon, dry tetrahydrofuran (THF, 100 mL) (Note 5) is added, and the flask is equipped with an internal thermometer. The solution is cooled to -100°C in a diethyl ether/liquid nitrogen bath and left for 5 min before slowly adding butyllithium (BuLi, 32 mL, 1.56 M in hexanes, 50 mmol) (approx. 20 min). After complete addition the mixture is stirred at -100°C for an additional 30 min before it is allowed to warm to -78°C. At this temperature, a solution of zinc bromide (ZnBr$_2$, 36.6 mL, 1.5 M in THF, 55 mmol) (Note 6) is slowly added (approx. 20 min). After complete addition, the reaction mixture is kept at -78°C for 5 min, then the flask is warmed with an ice bath to 0°C and left for 10 min at this temperature before allowing it to warm to room temperature. The yield of the zinc reagent is checked by hydrolysis and iodolysis (Note 7) before concentrating it under reduced pressure to 2.0-2.2 M (22-25 mL).

C. Ethyl 3-(4-cyanophenyl)propionate. A dry, 100-mL, three-necked flask, equipped with an argon inlet and a stirring bar, is charged with nickel acetylacetonate (Ni(acac)$_2$, 520 mg, 2 mmol) and evacuated for 10 min before flushing with argon. THF (6.7 mL), N-methylpyrrolidinone (NMP, 3.3 mL) (Note 8), 4-fluorostyrene (496 mg, 4 mmol) (Note 9) and ethyl 3-iodopropionate (4.56 g, 20 mmol) are successively added and the flask is equipped with an internal thermometer. The reaction mixture is cooled to -60°C before slowly adding the zinc reagent with a syringe through a large diameter cannula. After complete addition, the reaction mixture is allowed to warm to -14°C in a cryostat. (The checkers used a dry ice/ethylene glycol bath). The conversion is complete within 12-15 hr (Note 10), when it is quenched with saturated aqueous ammonium chloride solution (15 mL) and allowed to warm to room temperature. The quenched reaction mixture is extracted with diethyl ether (7 x 150 mL), the ethereal extracts are dried over magnesium sulfate, filtered, and evaporated to dryness by rotary evaporation at 40°C. The resulting yellow oil is purified by column chromatography (Note 11) affording 2.42 g (11.9 mmol) of ethyl 3-(4-cyanophenyl)propionate as a pale yellow oil (60%) (Note 12).

2. Notes

1. Ethyl 3-chloropropionate from Aldrich Chemical Company, Inc., was used as obtained.

2. Sodium iodide was purchased from Acros Organics as water free, 99+%.

3. Spectral data are as follows: IR (KBr) cm^{-1}: 2981 (m), 1372 (m), 1213 (s); ^{1}H NMR (300 MHz, CDCl$_3$) δ: 1.26 (t, 3 H, J = 7.1), 2.95 (t, 2 H, J = 7.5), 3.32 (t, 2 H, J = 7.5), 4.15 (q, 2 H, J = 7.1); ^{13}C NMR (75 MHz, CDCl$_3$) δ: -3.3, 14.6, 39.0,. 61.3, 171.4 MS (EI, 70 eV):), 228 (33), 183 (27), 155 (67), 101 (100), 73 (49). Anal. Calcd for C$_5$H$_9$IO$_2$: C, 26.34; H, 3.98. Found: C, 26.27; H, 3.96.

4. 4-Bromobenzonitrile from ABCR Germany is used as obtained.

5. THF is dried by distillation under argon from sodium/benzophenone.

6. Anhydrous ZnBr$_2$ is dried for 5 hr at 150°C under oil pump vacuum, then cooled to room temperature and flushed with argon before adding dry THF. The concentration is determined by transferring a 1-mL aliquot to a dry tared flask, then evaporating the THF.

7. Hydrolysis: An aliquot of the reaction mixture is quenched with saturated aqueous ammonium chloride solution and extracted with ether, then injected on GC to verify that all the 4-bromobenzonitrile has been consumed. Iodolysis: An aliquot of the reaction mixture is added to a dry vial containing iodine; after 10 min ether is added and the ethereal solution is washed with an aqueous solution of sodium thiosulfate. The organic phase is injected on

GC to verify the formation of the zinc reagent. Decane was used as internal standard in the reaction.

8.   NMP is dried by stirring overnight with calcium hydride, then refluxing for 5 hr followed by distillation under argon from calcium hydride.

9.   4-Fluorostyrene 99% from Aldrich Chemical Company, Inc., is used as obtained.

10.   The reaction is monitored by GC analysis of worked-up aliquots. Tetradecane is used as internal standard for the cross-coupling reaction.

11.   The oil is taken up in diethyl ether and absorbed onto approximately 15 g of flash silica (Merck silica gel 60 mesh 0.040-0.063 mm), then applied to a 10-cm diameter column packed with 500 g of flash silica eluting the product with pentane/diethyl ether 85:15. (The checkers used 10:1 hexanes/ether for improved resolution.)

12.   Spectral data are as follows: IR (KBr) cm^{-1}: 2983 (m), 2228 (m), 1732 (s), 1688 (m), 1186 (m). ^{1}H NMR (300 MHz, CDCl$_3$) δ: 1.20 (t, 3 H, J = 7.1), 2.60 (t, 2 H J = 7.5), 3.00 (t, 2 H, J = 7.5), 4.10 (q, 2 H, J = 7.1), 7.30 (d, 2 H, J = 7.8), 7.60 (d, 2 H, J = 7.8; ^{13}C NMR (75 MHz, CDCl$_3$) δ: 14.5, 31.3, 35.4, 61.0, 110.6, 119.2, 129.6, 132.6, 146.6, 172.5. MS (EI, 70 eV): 203 (26), 129 (100), 116 (39), 103 (12). Anal. Calcd. for C$_{12}$H$_{13}$NO$_2$: C, 70.92; H, 6.45; N, 6.89. Found: C, 70.61; H, 6.20; N, 6.74.

Waste Disposal Information

All toxic materials were disposed of in accordance with "Prudent Practices in the Laboratory"; National Academy Press; Washington, DC, 1995.

3. Discussion

The performance of cross-coupling reactions between aryl organometallics and alkyl iodides is not well known. Only the reaction of diarylcuprates with alkyl iodides may be considered for performing such cross-couplings.[2] The present procedure[3] describes a convenient way for performing the cross-coupling between an arylzinc bromide and an alkyl iodide. The reaction is catalyzed by Ni(acac)$_2$ (10 mol %) and the addition of commercially available 4-fluorostyrene (20 mol %). The role of 4-fluorostyrene is to reduce the electron density of the nickel intermediate [(Ar)Ni(Alkyl)] by coordinating to the nickel center and removing electron density, thereby favoring the reductive elimination leading to Ar-Alkyl. The key role of several electron-poor styrenes such as m- or p-trifluoromethylstyrene has been noticed in related cross-couplings between two Csp^3-centers.[4] The reaction tolerates a broad range of functional groups such as ester, nitrile, amide, and halogen (Table 1).

The functionalized arylzinc reagents are best prepared either starting from an aryllithium obtained by halogen-lithium exchange followed by a low-temperature (-80°C) transmetalation[5] with ZnBr$_2$ or by performing an iodine-magnesium exchange reaction. The latter reaction tolerates temperatures up to -30°C and is more convenient for industrial applications.[6]

1. Department of Chemistry, Ludwig-Maximilians-Universität München, Butenandtstr 5-13, D-81377 München, Germany. Knoch@cup.uni-muenchen.de

2. a) Lipshutz, B.H.; Sengupta, S. *Org. React.* **1992**, *41*, 135; see also: b) Erdik, E. *Tetrahedron* **1992**, *48*, 9577.

3. Giovannini, R.; Knochel, P. *J. Am. Chem. Soc.* **1998**, *120*, 11186.

4. Giovannini, R.; Stüdemann, T.;Dussin, G.; Knochel, P. Angew. *Chem., Int. Ed. Engl.* **1998**, *37,* 2387; b) Giovannini, R.; Stüdemann, T.; Devasagayaraj, A.; Dussin, G.; Knochel, P. *J. Org. Chem.* **1999**,*64*, 3544.

5. Tucker, C.E. Majid, T.N.; Knochel, P. *J. Am. Chem. Soc.* **1992**,*114*, 3983.

6. a) Boymond, L.; Rottländer, M.; Cahiez, G.; Knochel, P. Angew. *Chem., Int. Ed. Engl.***1998**, *37*, 1701; b) Abarbri, M.; Dehmel, F.; Knochel, P. *Tetrahedron Lett.* **1999**, *40*, 7449.

Table 1: Polyfunctional products obtained by the Ni(II)-catalyzed cross-coupling of Arylzinc bromides and alkyl iodides in the presence of 4-(trifluoromethyl)styrene.

Entry	ArZnBr	Alkyl iodide	Product	Yield (%)
1	Ph[b]	PivO(CH₂)₂I	PivO(CH₂)₂Ph	71
2	Ph[b]			75
3	p-Cl-C₆H₄[b]	EtO₂C(CH₂)₂I		71
4	p-MeO C₆H₄[b]	EtO₂C(CH₂)₂I		78
5	p-MeO C₆H₄[b]	BuCO(CH₂)₃I		72
6	p-MeO C₆H₄[b]			77
7	p-CN C₆H₄[b]	EtO₂C(CH₂)₂I		75
8	p-CN C₆H₄[b]			80
9	p-CN C₆H₄[b]			71
10	m-EtO₂C-C₆H₄[c]			72
11	m-EtO₂C-C₆H₄[c]	PhS(CH₂)₃I		75
12	o-EtO₂C-C₆H₄[c]			72

[a] Isolated yield of analytically pure products. [b] Prepared from the corresponding lithium reagent. [c] Prepared from the corresponding magnesium reagent.

Appendix

Chemical Abstracts Nomenclature (Collective Index Number);

(Registry Number)

Ethyl 3-(4-cyanophenyl)propionate: Benzenepropanoic acid, 4-cyano-, ethyl ester (12); (116460-89-0)

Ethyl 3-iodopropionate: Propanoic acid, 3-iodo-, ethyl ester (9); (6414-69-3)

Ethyl 3-chloropropionate: Propionic acid, 3-chloro-, ethyl ester (8); Propanoic acid, 3-chloro-, ethyl ester (9); (623-71-2)

Sodium iodide (8,9); (7681-82-5)

p-Cyanophenylzinc bromide: Zinc, bromo(4-cyanophenyl)- (12); (131379-14-1)

4-Bromobenzonitrile: Benzonitrile, p-bromo- (8); Benzonitrile, 4-bromo- (9); (623-00-7)

Butyllithium: Lithium, butyl- (8,9); (109-72-8)

Zinc bromide (8,9); (7699-45-8)

Nickel acetylacetonate: Nickel, bis(2,4-pentanedionato-) (8); Nickel, bis(2,4-pentanedionato-O, O')-, (sp-4-1)- (9); (3264-82-2)

N-Methylpyrrolidinone: 2-Pyrrolidinone, 1-methyl- (8,9); (872-50-4)

p-Fluorostyrene: Styrene, p-fluoro- (8); Benzene, 1-ethenyl-4-fluoro- (9); (405-99-2)

PREPARATION OF n-BUTYL 4-CHLOROPHENYL SULFIDE

A.

$$\text{(CF}_3\text{SO}_2)_2\text{O}$$
pyridine, DCM, -10°C → rt

B.

Pd(OAc)$_2$, NaHMDS
(R)-Tol-BINAP, toluene, 100°C

Submitted by J. Christopher McWilliams,[1] Fred J. Fleitz, Nan Zheng, and Joseph D. Armstrong III.

Checked by Scott E. Denmark and Ramzi F. Sweis

1. Procedure

A. 4-Chlorophenyl trifluoromethanesulfonate. A 250-mL, three-necked flask equipped with a Teflon-coated thermocouple, nitrogen bubbler, and septum is charged with a Teflon-coated magnetic stir bar, 4-chlorophenol (13.2 g, 102.5 mmol), and methylene chloride (125 mL) at room temperature (Note 1). Pyridine (9.1 mL, 112.5 mmol) is added to the solution via syringe, and the reaction mixture is cooled to -10°C with an ice/methanol cooling bath (Note 2). Triflic anhydride (18.7 mL, 111.3 mmol) is added dropwise to the reaction mixture via syringe at a rate such that the temperature remains below -2°C (Note 3). Upon completion of the addition, the reaction mixture is stirred for 1 hr at -10°C, then allowed to warm to room temperature. When the reaction is complete, water (75 mL) is added to the mixture and the resulting two-phase mixture is stirred for 15 min (Note 4). In a 500-mL separatory funnel, the layers are separated and the organic (lower) layer is washed sequentially with 0.2 N hydrochloric acid (HCl), water, and brine (75 mL each). The final organic layer is concentrated to a yellow oil via rotary evaporation. The yellow oil is diluted with 25 mL of 5% ethyl acetate in hexanes and filtered through a bed of silica gel (63 g charged to a 150-mL filter funnel and prewetted with 5% ethyl acetate in hexanes, Note 5). The silica gel is washed with 5% ethyl acetate in hexanes until the desired triflate is no longer detected in the filtrate (250 mL). The filtrate is concentrated via rotary evaporation to give the desired triflate as a clear, colorless liquid (25.7 g, 96% yield)(Note 6).

B. *n-Butyl 4-chlorophenyl sulfide.* A 1-L, single-necked, pear-shaped flask, equipped with a Teflon-coated magnetic stir bar and septum, is charged with 4-chlorophenyl trifluoromethanesulfonate (17.6 g, 67.5 mmol), palladium acetate [Pd(OAc)$_2$, (0.9 g, 4.1 mmol)], and toluene (425 mL), (Note 7). A nitrogen source is introduced through a 12-in., 16-gauge needle, which is punctured through the septum and submerged into the mixture. A second 16-gauge needle leading to a bubbler is punctured through the septum. The mixture is stirred while vigorously purging with nitrogen for 15 min (Note 8). The septum is opened briefly to introduce [(R)-Tol-BINAP, (3.1 g, 4.5 mmol, Note 9)], and the mixture is stirred for 15 min with continuous purging with nitrogen, yielding a homogenous orange solution.

A 2-L, three-necked, round-bottomed flask is equipped with a mechanical stirrer, reflux condenser leading to a nitrogen inlet and a septum. A Teflon-coated thermocouple is introduced into the flask through the septum. The septum is removed and the flask is charged with toluene (515 mL). A second nitrogen source is introduced into the solution through a 12-in., 16-gauge needle, and the solution is vigorously purged with nitrogen for 5 min.   To this mixture is immediately added via syringe 0.6 M sodium bis(trimethylsilyl)amide in toluene (157.5 mL, 94.5 mmol, Note 10). While the solution is stirred vigorously, 1-butanethiol (10.1 mL, 94.5 mmol) is added over the course of 3 min (Note 11). A gel-like solid forms and the internal temperature increases from 21°C to 28°C.

44

The solution containing the palladium catalyst and 4-chlorophenyl trifluoromethanesulfonate in toluene is transferred via a cannula to the mixture containing sodium 1-butanethiolate (Note 12). The mixture is heated to 100°C and stirred at this temperature for 12 hr (Note 13). The mixture is cooled to ambient temperature and transferred to a 2-L extraction funnel. The organic phase is washed with 2 N sodium hydroxide (125 mL), 2 N hydrochloric acid (100 mL) and an aqueous saturated solution of sodium chloride (50 mL). The organic phase is dried over sodium sulfate and concentrated on a rotary evaporator, yielding a heterogeneous mixture. The mixture is triturated with hexanes (150 mL) and stored at 0°C overnight (Note 14). The mixture is filtered through a sintered-glass funnel, washing with hexanes. The filtrate is concentrated on a rotary evaporator, and the concentrate is transferred to a 35-mL, pear-shaped flask equipped with a Teflon-coated magnetic stir bar, a 12-cm Vigreux column and a short-path distillation apparatus. The oil is distilled at reduced pressure (0.3 mm), discarding the fractions that boil at 55-91°C (Note 15). The fraction boiling at 92°C is collected, to yield n-butyl 4-chlorophenyl sulfide as a colorless oil (11.2 g, 83% yield)(Notes 16,17).

2. Notes

1. An atmosphere of nitrogen is maintained throughout the course of the reaction. 4-Chlorophenol and dichloromethane were purchased from Aldrich Chemical Company, Inc., and used without further purification.

2. Unless otherwise noted, all temperatures refer to internal temperatures measured with Teflon-coated thermocouples. Pyridine was purchased from Aldrich Chemical Company, Inc., and used without further purification.

3. Triflic anhydride was purchased from Aldrich Chemical Company, Inc., and used without further purification.

4. The course of the reaction was followed by TLC (95:5 ethyl acetate/hexanes, Silica Gel 60 F254, 2.5 x 7.5 cm, 250 μ thickness, UV visualization at 254 nm), monitoring for the disappearance of 4-chlorophenol: TLC: R_f (4-chlorophenol) = 0.09, R_f (4-chlorophenyl trifluoromethanesulfonate) = 0.56.

5. Silica Gel 60 (230-400 mesh) was purchased from EM Science.

6. The physical properties are as follows: ^{1}H NMR (400 MHz, CDCl$_3$) δ: 7.24 (m, 2 H, J = 9.0, 3.4, 2.2), 7.44 (m, 2 H, J = 8.9, 3.4, 2.1); ^{13}C NMR (100 MHz, CDCl$_3$) δ: 118.7 (q, J = 316.8), 122.6, 130.2, 134.2, 147.8, ppm; IR (neat) cm^{-1}: 1208, 1139, 880. Anal. Calcd for C$_7$H$_4$ClF$_3$O$_3$S: C, 32.26; H, 1.55. Found: C, 32.25; H, 1.65.

7. Pd(OAc)$_2$ was purchased from Strem Chemical Co. and toluene was purchased from EM Science.

8. Most of the solids dissolve within 15 min.

9. (R)-Tol-BINAP [(*R*)-(+)-2,2'-bis(di-p-tolylphosphino)-1,1'-binaphthyl] was purchased from Strem Chemical Co., and used without further purification.

10. The 0.6 M solution of sodium bis(trimethylsilyl)amide in toluene was purchased from Aldrich Chemical Company, Inc., and used as received.

11. 1-Butanethiol was purchased from Aldrich Chemical Company, Inc., and stored over 4Å molecular sieves. An excess of thiol is employed because of the formation of the disulfide as a competing side reaction.

12. The color of the mixture becomes bronze.

13. The submitters followed the course of the reaction by HPLC, monitoring for the disappearance of 4-chlorophenyl trifluoromethanesulfonate. HPLC conditions were as follows: 0.46 x 7.5 cm Eclipse-XDB column, gradient elution (10:80 → 85:15 acetonitrile/10 mM pH 6.3 phosphate buffer over 8.5 min, flow rate = 1.5 mL/min, detection = 210 nm). R_T (4-chlorophenyl trifluoromethanesulfonate) = 6.25 min, R_T (n-butyl 4-chlorophenyl sulfide) = 7.10 min.

46

14. Triturating the mixture with hexanes and storing overnight at 0°C induces the precipitation of solid (R)-Tol-BINAP, which is then removed by filtration.

15. Distillation is required to remove the primary side-product, di-n-butyl disulfide. Purification by silica gel chromatography gave inferior results because of similar retention times of the two compounds. The initial distillate fractions are turbid, indicating the presence of di-n-butyl disulfide. The absence of turbidity in the higher boiling fraction indicates distillate that is predominantly the desired n-butyl 4-chlorophenyl sulfide, containing only small amounts of n-butyl disulfide.

16. The isolated oil contains 0.5 mol% n-butyl disulfide. Additional n-butyl 4-chlorophenyl sulfide remains in the distilling flask, but is not recovered. The physical properties are as follows: ^{1}H NMR (500 MHz, CDCl$_3$) δ: 0.94 (t, 3 H, J = 7.3), 1.43-1.48 (m, 2 H), 1.59-1.65 (m, 2 H), 2.90 (t, 2 H, J = 14.6), 7.27 (br s, 4 H); ^{13}C NMR (126 MHz, CDCl$_3$) δ: 13.6, 21.9, 31.1, 33.5, 128.9, 130.2, 131.6, 135.6; IR (neat) cm^{-1}: 1096, 810. Anal. Calcd for C$_{10}$H$_{13}$ClS: C, 59.84; H, 6.53. Found: C, 59.62; H, 6.76.

17. The checkers employed rac-BINAP purchased from Strem Chemical Company and used it in place of (R)-Tol-BINAP.[3] The same procedure was followed, except that 3 equiv of lithium chloride was added to the complete reaction mixture immediately prior to heating to 100°C. The reaction was run on one-half the scale used for (R)-Tol-BINAP and the product was obtained as before after two distillations (5.07 g, 75% yield). Anal. Calcd for C$_{10}$H$_{13}$ClS: C, 59.84; H, 6.53. Found: C, 59.85; H, 6.75. The initial dissolution of rac-BINAP is not complete because of the lower solubility of this ligand compared with (R)-Tol-BINAP.

Waste Disposal Information

All toxic materials were disposed of in accordance with "Prudent Practices in the Laboratory"; National Academy Press; Washington, DC, 1995.

3. Discussion

This procedure describes the preparation of n-butyl 4-chlorophenyl sulfide starting from 4-chlorophenol. The intermediate 4-chlorophenyl trifluoromethanesulfonate highlights the chemoselectivity of this procedure, which overwhelmingly favors coupling at the carbon bearing the triflate functionality. As a consequence, the product contains a handle for additional functionalization by any one of the newly developed methods for coupling nucleophiles to aryl chlorides.[2] This procedure also highlights a modification of the originally described protocol, in which sodium *tert*-butoxide had been employed as the base.[3] Since the submitters' initial disclosure, they have discovered that the use of sodium bis(trimethylsilyl)amide decreases the amount of phenol by-product formed during the course of the reaction. In head-to-head experiments with sodium *tert*-butoxide versus bis(trimethylsilyl)amide, greater amounts of phenol were observed with the former base. This recent development allows the use of higher reaction temperatures and shorter reaction times if desired. The same yield of n-butyl 4-chlorophenyl sulfide was obtained when the reaction was conducted at 80°C, but a longer reaction time was required. Furthermore, the commercially available solutions of bis(trimethylsilyl)amide in toluene can be used directly, obviating the usual difficulties encountered with handling hygroscopic solids.

The preparations of aryl sulfides typically employ aryl halides as starting materials.[4] The procedure described here makes use of the ubiquitous class of commercially available phenolic compounds in the form of aryl triflates, which expands the range of readily accessible aryl sulfides. Prior to this disclosure, the use of aryl triflates in a palladium-catalyzed process for the formation of aryl alkyl sulfides was unprecedented.[5] This procedure appears to be general with regard to electronically neutral or electron-deficient aryl triflates (Table 1). The yields in Table I correspond to the initially disclosed procedure employing sodium *tert*-butoxide as the base. Lower yields were obtained with the 4-nitro-substituted triflate, and in the case of the electronically enriched substrate containing a 4-methoxy group.

1. Merck & Co., Inc., P.O. Box 2000, RY800-C362, Rahway, NJ 07065.

2. For examples, see: (a) Littke, A. F.; Fu, G. C. *J. Am. Chem. Soc.* **2001**, *123*, 6989; (b) Li, G. Y.; Zheng, G.; Noonan, A. F. *J. Org. Chem.* **2001**, *66*, 8677; (c) Lee, H. M.; Nolan, S. P. *Org. Lett.* **2000**, *2*, 2053.

3. Zheng, N.; McWilliams, J. C.; Fleitz, F. J.; Armstrong, III, J. D.; Volante, R. P. *J. Org. Chem.* **1998**, *63*, 9606.

4. (a) For recent examples, see reference 2b and additional references sited therein; (b) For a general review on sulfides, see: "Comprehensive Organic Chemistry"; Jones, D. N., Ed; Pergamon Press: Oxford, 1979; Vol. 3, pp 33-103; (c) For a review of transition-metal catalyzed carbon-heteroatom bond formation, see: Baranano, D.; Mann, G.; Hartwig, J. F. *Curr. Org. Chem.* **1997**, *1*, 287.

5. (a) A nickel-catalyzed process employing aryl mesylates and thiophenols has been described: Percec, V.; Bae, J.-Y.; Hill, D. H. *J. Org. Chem.* **1995**, *60*, 6895; (b) For the preparation of aryl silyl sulfides from aryl triflates and trialkylsilanethiolates, see: Arnould, J. C.; Didelot, M.; Cadilhac, C.; Pasquet, M. J. *Tetrahedron Lett.* **1996**, *37*, 4523.

TABLE 1. PREPARATION OF ALKYL ARYL THIOEHERS

Aryl Triflate	Thiol	Product	Yield (%)
Ph–OTf	HS–CH2CH2CH3	Ph–S–CH2CH2CH2CH3	93
Ph–OTf	HS–C(CH3)3	Ph–S–C(CH3)3	66
2-methylphenyl–OTf	HS–CH2CH2CH3	2-methylphenyl–S–CH2CH2CH2CH3	79
4-tert-butylphenyl–OTf	HS–CH2CH2CH3	4-tert-butylphenyl–S–CH2CH2CH2CH3	92
4-MeO–phenyl–OTf	HS–CH2CH2CH3	4-MeO–phenyl–S–CH2CH2CH2CH3	54
4-NC–phenyl–OTf	HS–CH2CH2CH3	4-NC–phenyl–S–CH2CH2CH2CH3	82
4-O2N–phenyl–OTf	HS–CH2CH2CH3	4-O2N–phenyl–S–CH2CH2CH2CH3	60
4-PhC(O)–phenyl–OTf	HS–CH2CH2CH3	4-PhC(O)–phenyl–S–CH2CH2CH2CH3	91
1-naphthyl–OTf	HS–CH2CH2CH3	1-naphthyl–S–CH2CH2CH2CH3	95

50

Appendix

Chemical Abstracts Nomenclature (Collective Index Number);

(Registry Number)

4-Chlorophenyl trifluoromethanesulfonate: Methanesulfonic acid, trifluoro-, p-chlorophenyl ester (8); Methanesulfonic acid, trifluoro-, 4-chlorophenyl ester (9); (29540-84-9)

4-Chlorophenol: TOXIC: Phenol, p-chloro- (8); Phenol, 4-chloro- (9); (106-48-9)

Pyridine (8, 9); (110-86-1)

Triflic anhydride: Methanesulfonic acid, trifluoro-, anhydride (8, 9); (358-23-6)

Palladium acetate: Acetic acid, palladium(2+) salt (8, 9); (3375-31-3)

(R)-(+)-2,2'-Bis(di-p-tolylphosphino)-1,1'-binaphthyl: STREM CHEMICALS: (R)-Tol-BINAP: Phosphine, [1,1'-binaphthalene]-2,2'-diylbis[bis(4-methylphenyl)-, (R)- (11); (99646-28-3)

Sodium bis(trimethylsilyl)amide: Disilazane, 1,1,1,3,3,3-hexamethyl-, sodium salt (8); Silanamine, 1,1,1-trimethyl-N-(trimethylsilyl)-, sodium salt (9); (1070-89-9)

1-Butanethiol (8, 9); (109-79-5)

rac-2,2'-Bis(diphenylphosphino)-1,1'-binaphthyl: rac-BINAP: Phosphine, [1,1'-binaphthalene]-2,2'-diylbis[diphenyl]- (11); (98327-87-8)

SYNTHESIS OF SYMMETRICAL trans-STILBENES BY A DOUBLE HECK REACTION OF (ARYLAZO)AMINES WITH VINYLTRIETHOXYSILANE: trans-4,4'-DIBROMOSTILBENE

{Benzene, 1,1'-(1,2-ethenediyl)bis[4-bromo-, (E)- using Silane, ethenyltriethoxy-}

Submitted by Saumitra Sengupta and Subir K. Sadhukhan.[1]
Checked by Kimberly Savary and Edward J. J. Grabowski.

1. Procedure

A. *4-[(Bromophenyl)azo]morpholine.* In a 500-mL Erlenmeyer flask equipped with a magnetic stirring bar are placed 4-bromoaniline (15.0 g, 87 mmol, Note 1) and 6 N hydrochloric acid (HCl), 36.4 mL, 210 mmol) and the mixture is warmed on a water bath to make a clear solution. It is cooled to 0°C to produce a heavy precipitate. A solution of sodium nitrite (6.30 g, 91 mmol) in water (10 mL) is added dropwise over 10 min. Stirring is continued at 0°C for 20 min (Note 2), and morpholine (8.3 g, 9.0 mL, 96 mmol) is added dropwise to the above solution over 10 min. Water (100 mL) is added followed by the dropwise addition of 10% aqueous sodium bicarbonate solution (130 mL) (Note 3). After the solution is stirred for a further hour, the precipitated solid is filtered, washed with water and dried in air. The solid is dissolved in hot light petroleum (60-80 fraction) (80 mL) and treated with activated charcoal (1.5 g). The mixture is filtered while hot and the filtrate concentrated to ca. 40 mL. Upon cooling to room temperature, shiny crystals of the pure triazene are obtained. The mother liquor is concentrated to give a second crop of crystals (combined yield 20.3 g, 85%, Notes 4, 5).

B. *trans-4,4'-Dibromostilbene*. A 500-mL round-bottomed flask equipped with a magnetic stirring bar is charged with the above triazene (14.3 g, 53 mmol) and methanol (125 mL). The stirred solution is cooled to 0°C and 40% tetrafluoroboric acid (HBF$_4$, 23 mL, 106 mmol) is added dropwise over 10 min. After the addition is complete, the ice bath is removed and the reaction brought to room temperature. It is stirred for an additional 10 min (Note 6) and palladium acetate [Pd(OAc)$_2$, 0.12 g, 0.53 mmol] is added followed by the dropwise addition of a solution of vinyltriethoxysilane (4.94 g, 5.5 mL, 26 mmol) in methanol (10 mL). A second lot of Pd(OAc)$_2$ (0.12 g, 0.53 mmol) is added and stirring continued for a further 30 min at room temperature (Note 7). The mixture is warmed to 40°C for 20 min and finally heated under reflux for 15 min (Note 8). The solution is concentrated under reduced pressure to half its volume and water (150 mL) is added. The precipitated solid is filtered, washed with water and dried in air. It is then boiled with toluene (125 mL) and filtered while hot. The filtrate is concentrated to ca. 70 mL, warmed to 70°C and light petroleum (30 mL) is added. Upon cooling to room temperature, the product crystallizes (4.20 g). Concentration of the mother liquor gives an additional crop (0.70 g) of the product (combined yield 4.90 g, 46.5%, Note 9).

2. Notes

1. 4-Bromoaniline (Lancaster Laboratories), morpholine (Lancaster Laboratories),

vinyltriethoxysilane (Lancaster Laboratories), palladium acetate (Arora-Mathey) and 40% HBF_4 (Central Drug House) were used as received.

2. At this point a clear solution is obtained.

3. *Caution:* Vigorous evolution of carbon dioxide occurs. The product began to precipitate. The checkers used 1 N sodium hydroxide (NaOH, approximately 120 mL to pH~7 at 0-10°C) to effect neutralization and avoid difficulties with carbon dioxide evolution.

4. *Caution:* All 1-aryl triazenes are toxic and direct hand contact should be avoided. The checkers used hexane for the recrystallization.

5. The product showed the following physical data: mp 87-88°C (lit.[2] mp 89.5-90°C); IR (CHCl$_3$) cm^{-1}: 1100, 1210, 1345, 1430, 1480, 3010; ^{1}H NMR (300 MHz, CDCl$_3$) δ: 3.76-3.87 (AA'BB', 8 H), 7.29-7.48 (AA'BB', 4 H); Anal. Calcd for $C_{10}H_{12}BrN_3O$: C, 44.47; H,4.44; N,15.56. Found: C, 44.22; H, 4.38; N, 15.50.

6. The solution became clear at this stage.

7. Some of the product, as it formed, precipitated.

8. Inferior yields were obtained when the reaction is heated directly under reflux without being held at 40°C for ca. 20 min.

9. The product showed the following physical data: mp 214–215°C (lit.[3] m p 215–216°C); ^{1}H NMR (300 MHz, CDCl$_3$) δ: 7.01 (s, 2 H), 7.35 (d, 4 H, J = 6.9), 7.47 (d, 4 H, J = 6.9).

Waste Disposal Information

All toxic materials were disposed of in accordance with "Prudent Practices in the Laboratory"; National Academy Press; Washington, DC, 1995.

3. Discussion

Stilbenes are an important class of chromophores that have found various applications in molecular photonics and optoelectronics.[4] Symmetrical trans-stilbenes are classically prepared via oxidative dimerization of benzylic halides, which usually require strongly basic conditions and high reaction temperatures.[5] More recent methods include McMurry-coupling of aromatic aldehydes,[6] Wittig condensations and double Heck reactions of haloarenes with ethylene.[7] The latter procedure, however, is inconvenient since it requires a measured amount (half equivalent) of ethylene gas under autoclave conditions. Moreover, a double Heck reaction of aryl bromides and iodides with ethylene invariably produces some of the undesired 1,1-diarylethylene regioisomer (up to 20%). The stilbene synthesis described here is based on the submitters' recent report on double Heck reactions of arenediazonium salts with vinyltriethoxysilane (Scheme 1).[8] The procedure has been suitably modified so that the arenediazonium salt is generated in situ from 1-aryltriazene, thus avoiding direct handling of the diazonium salt.[9]

Scheme 1

$$R = p\text{-Me} \ (56\%)$$
$$p\text{-Cl} \ (55\%)$$
$$p\text{-Br} \ (65\%)$$
$$p\text{-I} \ (60\%)$$
$$p\text{-OMe} \ (45\%)$$
$$p\text{-NO}_2 \ (67\%)$$
$$o\text{-Me} \ (46\%)$$
$$o\text{-CO}_2\text{Me} \ (68\%)$$

Synthesis of symmetrical stilbenes via a double Heck reaction of arenediazonium salts with vinyltriethoxysilane is attractive on several counts: Ready availability of starting materials (anilines are more readily available in a diverse substitution pattern than are aryl halides and aldehydes); Mild reaction conditions that tolerate a number of functional groups (e.g., NO_2, CO_2R); Strategic use of vinyltriethoxysilane as a cheap and easily handled ethylene equivalent; And operational simplicity (alcoholic reaction medium, ligandless Pd-catalyst, no additives). Most significantly, by virtue of the superior Heck reactivity of the diazonium nucleofuge over bromide (and even iodides),[10,11] the present methodology allows for a facile synthesis of 4,4'-dibromo- (and 4,4'-diiodo-) stilbenes, which would otherwise be difficult to synthesize (impossible for the diiodo derivative) via the double Heck reaction of haloarenes with ethylene. Recently, the submitters have also described the double Heck reaction of bisarenediazonium salts with vinyltriethoxysilane for the synthesis of poly(1,4-biarylenevinylenes).[12]

The dibromo- (and diiodo-) stilbenes are potentially useful monomers for the synthesis of poly(p-phenylenevinylenes) via poly-Heck or poly-Suzuki coupling reactions.[13]

1. Department of Chemistry, Jadavpur University, Calcutta 700 032, India.

2. Henry, R. A.; Dehn, W. M. *J. Am. Chem. Soc.* **1943**, *65*, 479.

3. (a) Cadogan, J. I. G.; Inward, P. W. *J. Chem. Soc.* **1962**, 4170; (b) Mataka, S.; Liu, G. -B.; Tashiro, M. *Synthesis* **1995**, 133.

4. (a) Kraft, A.; Grimsdale, A. C.; Holmes, A. B. *Angew. Chem., Int. Ed. Engl.* **1998**, *37*, 402; (b) Meier, H.; Lehmann, M. *Angew. Chem., Int. Ed. Engl.* **1998**, *37*, 643.

5. Becker, K. B. *Synthesis* **1983**, 341.

6. McMurry, J. E. *Chem. Rev.* **1989**, *89*, 1513.

7. Klingelhofer, S.; Schellenberg, C.; Pommerehne, J.; Bassler, H.; Greiner, A.; Heitz, W. *Macromol. Chem. Phys.* **1997**, *198*, 1511.

8. Sengupta, S.; Bhattacharya, S.; Sadhukhan, S. K. *J. Chem. Soc., Perkin Trans. 1* **1998**, 275.

9. (a) Bhattacharya, S.; Majee, S.; Mukherjee, R.; Sengupta, S. *Synth. Commun.* **1995**, *25*, 651; (b) Sengupta, S.; Sadhukhan, S. K.; Bhattacharyya, S. *Tetrahedron* **1997**, *53*, 2213.

10. Beletskaya, I. P.; Cheprakov, A. V. *Chem. Rev.* **2000**, *100*, 3009.

11. (a) Kikukawa, K.; Nagira, K.; Wada, F.; Matsuda, T. *Tetrahedron* **1981**, *37*, 31; (b) Sengupta, S.; Bhattacharya, S. *J. Chem. Soc., Perkin Trans. 1* **1993**, 1943; (c) Sengupta, S.; Sadhukhan, S. K. *Tetrahedron Lett.* **1998**, *39*, 715.

12. Sengupta, S.; Sadhukhan, S. K. *J. Chem. Soc., Perkin Trans. 1* **1999**, 2235.

13. Reviews: (a) Scherf, U.; Müllen, K. *Synthesis* **1992**, 23; (b) de Meijere, A.; Meyer, F. E. *Angew. Chem., Int. Ed. Engl.* **1994**, *33*, 2379; (c) Heitz, W. *Pure Appl. Chem.* **1995**, *67*, 1951; (d) Greiner, A.; Bolle, B.; Hesemann, P.; Oberski, J. M.; Sander, R. *Macromol. Chem. Phys.* **1996**, *197*, 113; (e) Herrmann, W. A. In "Applied Homogeneuos Catalysis with Organometallic Compounds"; Cornils, B., Herrmann, W. A., Eds.; VCH: Weinheim, 1996; Ch. 3.1.6, p. 712.

Appendix

Chemical Abstracts Nomenclature (Collective Index Number);

(Registry Number)

Vinyltriethoxysilane: ALDRICH: Triethoxyvinylsilane:

Silane, triethoxyvinyl- (8); Silane, ethenyltriethoxy- (9); (78-08-0)

trans-4,4'-Dibromostilbene: Stilbene, 4,4'-dibromo-, (*E*)- (8); Benzene, 1,1'-(1,2-ethenediyl)bis[4-bromo-, (*E*)- (9); (18869-30-2)

4-Bromoaniline: Aniline‿p-bromo- (8); Benzenamine, 4-bromo- (9); (106-40-1)

4-[(4-Bromophenyl)azo]morpholine: Morpholine, 4-[(4-bromophenyl)azo]- (14); (188289-57-8)

Sodium nitrite: Nitrous acid, sodium salt (8, 9); (7632-00-0)

Morpholine (8, 9); (110-91-8)

Tetrafluoroboric acid: Borate (1-), tetrafluoro-, hydrogen (8, 9); (16872-11-0)

Palladium acetate: Acetic acid, palladium(2+) salt (8, 9); (3375-31-3)

SYNTHESIS AND UTILIZATION OF INDIUM (I) IODIDE FOR IN SITU FORMATION OF ENANTIOENRICHED ALLENYLINDIUM REAGENTS AND THEIR ADDITION TO ALDEHYDES: (2R,3S,4S)-1-(tert-BUTYLDIPHENYLSILYLOXY)-2,4-DIMETHYL-5-HEXYN-3-OL

[5-Hexyn-3-ol, 1-[[(1,1-dimethylethyl)diphenylsilyl]oxy]-2,4-dimethyl-, (2R,3S,4S)-]

A. $2 \text{ In} + 3 \text{ I}_2 \xrightarrow[\Delta]{\text{xylenes}} 2 \text{ InI}_3$

B. $4 \text{ InI}_3 + 2 \text{ In} \xrightarrow[\text{2) Et}_2\text{O}]{\text{1) xylenes, } \Delta} 3 \text{ InI} + 3 \text{ InI}_3 \cdot \text{OEt}_2$

C.

D.

E.

Submitted by Brian A. Johns, Charsetta M. Grant, and James A. Marshall.[1]
Checked by Edward B. Holson and William R. Roush

1. Procedure

A. Indium(III) iodide. To a 1-L, oven-dried, round-bottomed flask flushed with argon and equipped with a magnetic stirrer is added xylenes (500 mL). The solvent is degassed

(Note 1) and the flask is equipped with a reflux condenser. Indium powder (5.00 g, 43.6 mmol) (Note 2) is added to the vigorously stirring solution, followed by iodine (I_2, 16.57 g, 65.32 mmol). The mixture is stirred vigorously (Note 3) and heated at reflux (bath temp. ~160-170°C) under argon for 1-1.5 hr or until the indium metal is consumed. When the metal is consumed, a crystal of iodine is added and stirring at reflux is resumed. The reaction is complete when the added iodine is not consumed after 15 min at reflux. The solution is filtered hot and allowed to cool to room temperature (Note 4). The resulting bright yellow crystals are filtered under nitrogen using a Schlenk filtration system (Note 5) and washed with two 10-mL portions of cold benzene to remove traces of I_2. The filtrate is concentrated to 1/4-1/3 volume (Note 6) and cooled to 0°C. The yellow precipitate is Schlenk-filtered under nitrogen and washed with cold benzene (10 mL). The product is dried under reduced pressure to yield 18.3 g (85%) of indium(III) iodide [In(III)I] (Note 7).

B. *Indium(I) iodide.* To a 1-L, oven-dried, round-bottomed flask flushed with argon and equipped with a magnetic stirrer is added xylenes (400 mL). The solvent is degassed (Note 1) and the flask is equipped with a reflux condenser. Indium(III) iodide (18.30 g, 36.93 mmol) is added to the flask, and the mixture is stirred vigorously while indium powder (2.12 g, 18.46 mmol) (Note 2) is added. The mixture is stirred vigorously (Note 3) at reflux under argon for 18 hr. The resulting yellow suspension is allowed to cool to room temperature, diluted with ether (400-500 mL) and stirred for 1 hr. The resulting burgundy precipitate is filtered under air and washed with ether (100 mL). The product is dried under reduced pressure to yield 6.14 g (92%) of indium(I) iodide. The filtrate is concentrated to dryness on a rotary evaporator, venting with nitrogen, to yield 14.4 g (105% based on eq B.) (Note 8) of recovered indium(III) iodide (Note 9).

C. *(R)-3-(tert-Butyldiphenylsilyloxy)-2-methylpropanal.* A 500-mL, oven-dried, round-bottomed flask equipped with a magnetic stirrer is charged with 10.00 g (84.67 mmol) of methyl (R)-(-)-3-hydroxy-2-methylpropionate (Note 10), and 100 mL of N,N-dimethylformamide (Note 11). The solution is cooled to 0°C and 14.4 g (211 mmol) of

60

imidazole is added. Upon dissolution of the imidazole, 23.1 mL (88.9 mmol) of tert-butyldiphenylchlorosilane (DPSCl) (Note 12) is added dropwise via syringe. After 10 min, the cooling bath is removed and the solution is allowed to warm to room temperature for 2 hr. The reaction is quenched by the addition of 200 mL of pentane and 40 mL of water (H_2O). The layers are separated and the pentane layer is washed with brine (75 mL). The aqueous layer is extracted with pentane (3 x 75 mL), and the combined extracts are dried over anhydrous sodium sulfate (Na_2SO_4). Filtration and concentration under reduced pressure followed by purification by flash chromatography (Note 13) yield 29.9 g (99%) of methyl (R)-3-(tert-butyldiphenylsilyloxy)-2-methylpropionate as a clear oil (Note 14).

A 1-L, oven-dried, round-bottomed flask equipped with a magnetic stirrer is charged with 9.92 g (27.9 mmol) of methyl (R)-3-(tert-butyldiphenylsilyloxy)-2-methylpropionate and 200 mL of dry hexanes (Note 15). The solution is cooled to -78°C, and 31.5 mL (31.5 mmol) of 1 M diisobutylaluminum hydride (in hexane) (DIBAL-H) (Note 16) is added dropwise over 15 min via a syringe pump. After the addition is complete, the resultant solution is stirred at -78°C for 2 hr. The reaction is quenched by pouring the cold solution into 250 mL of saturated aqueous Rochelle's salt. Ether (300 mL) and H_2O (75 mL) are added and the biphasic mixture is stirred vigorously for 1 hr (Note 17). The layers are separated and the ether layer is washed with brine. The aqueous layer is extracted with ether (2 x 50 mL) and the combined extracts are dried over Na_2SO_4. Filtration of the solution and concentration of the filtrate under reduced pressure followed by purification of the crude product by flash chromatography (Note 18) yields 7.85 g (86%) of (R)-3-(tert-butyldiphenylsilyloxy)-2-methylpropanal as a white solid (Note 19).

D. (R)-3-Butyn-2-yl methanesulfonate. To a 1-L, oven-dried, round-bottomed flask flushed with argon and equipped with a magnetic stirrer is added dichloromethane (CH_2Cl_2, 713 mL) and (R)-(+)-3-butyn-2-ol (10.00 g, 143 mmol) (Note 20). The mixture is cooled to -78°C and triethylamine (Et_3N, 39.7 mL, 285 mmol) and methanesulfonyl chloride (16.6 mL, 214 mmol) are added. The resulting mixture is stirred at -78°C for 1 hr, then quenched with aqueous saturated sodium bicarbonate ($NaHCO_3$) solution and allowed to warm to room temperature. The layers are separated and the organic layer is washed with brine and concentrated under aspirator pressure (Note 21). The residue is diluted with 500 mL of ether and washed with water (20 mL) followed by brine (20 mL). The aqueous layer is extracted with ether (50 mL). The combined extracts are dried over anhydrous Na_2SO_4 and concentrated under aspirator pressure to yield 20.13 g (95%) of the methanesulfonate (Note 22). The material is used without further purification (Note 23).

E.   *(2R,3S,4S)-1-(tert-Butyldiphenylsilyloxy)-2,4-dimethyl-5-hexyn-3-ol.*   An oven-dried, 100-mL, one-necked flask, equipped with a magnetic stirring bar is purged with argon. The flask is charged with (R)-3-(tert-butyldiphenylsilyloxy)-2-methylpropanal (3.00 g, 9.19 mmol) and the (R)-methanesulfonate (1.50 g, 10.1 mmol). Tetrahydrofuran (THF, 29.4 mL) and hexamethylphosphoramide (HMPA, 7.4 mL) are added via syringe. To the solution is added $PdCl_2(dppf)$ (335 mg, 0.46 mmol) (Note 24), immediately followed by indium(I) iodide (2.66 g, 11.0 mmol). The resultant dark suspension is stirred vigorously for 1 hr at which time the reaction is judged complete by TLC (Note 25). The reaction mixture is quenched by the addition of H_2O (30 mL), and ether (20 mL) is added. After being stirred for 2 min the layers are separated and the ether layer is washed with brine. The aqueous layer is extracted with ether (3 x 25 mL) and the combined extracts are dried over anhydrous Na_2SO_4. Filtration of the solution and concentration of the filtrate under reduced pressure followed by purification of the crude product by flash chromatography on silica gel (Note 26) provide 2.64 g (76%) of (2R,3S,4S)-1-(tert-butyldiphenylsilyloxy)-2,4-dimethyl-5-hexyn-3-ol (Note 27) as a clear oil and 201 mg of the (2R,3R,4R) diastereomer (Note 28).

2. Notes

1. The commercial mixture of xylenes was used as received without further

purification. Degassing was accomplished by bubbling argon through the solvent for 30 min.

2. Indium powder, -100 mesh, 99.99%, was purchased from Aldrich Chemical Company, Inc.

3. Vigorous stirring is very important. Use of a large stir bar for this reaction is crucial. During the course of the reaction the powdered indium metal may adhere to the sides of the flask. On occasion it was necessary to dislodge this powder from the walls of the flask. The flask was removed from the heat source, allowed to cool below reflux, and the walls of the flask were scraped with a metal spatula. Caution should be exercised when performing this operation to avoid burns and the flask should be blanketed with argon to prevent solvent flash. On a larger scale the submitters recommend use of a mechanical stirrer.

4. A coarse, 150-mL, fritted glass filter was used for the hot filtration. This filtration should be performed quickly, as precipitates form rapidly upon cooling. This filtration may be performed in the air without complication.

5. The product is very hygroscopic and should be handled under an inert atmosphere. The checkers found that inactive material was produced if the filtration was performed using a 150-mL fritted funnel under a blanket of nitrogen. The checkers subsequently performed this filtration under nitrogen using a Schlenk filtration system consisting of a 2-L, 24/40 three-necked flask, a 24/40 double male adapter, and a 400-mL, 24/40, coarse-fritted Schlenk filter. The reaction vessel is connected to one end of the double male adapter and the other end is quickly attached to the Schlenk filter system under a positive flow of nitrogen. The solution is poured onto the filter, and filtered into the three-necked flask via a vacuum system that is connected to one of the ports on the three-necked flask.

6. This step may be performed by using a rotary evaporator connected to a water aspirator, with a drying column inserted between the aspirator and the evaporator. The rotary evaporator is vented with nitrogen once the desired final volume is reached.

7. The checkers obtained 74-85% yields of indium(III) iodide.

8. The checkers obtained 86-93% yields of indium(I) iodide, and 98-113% yields of recovered indium(III) iodide etherate. The latter material was successfully recycled as described by the submitters (Note 9).

9. The recovered InI_3 can be used without further purification to generate additional InI. Procedure B was repeated with recovered InI_3 (14.4 g, 29.1 mmol), indium powder (1.7 g, 14.8 mmol) and degassed xylenes (320 mL) to yield 5.0 g (95%) of indium(I) iodide and 10.0 g (93%) of recovered indium(III) iodide.

10. Methyl (R)-(-)-3-hydroxy-2-methylpropionate was purchased from Sigma Chemical Company and refrigerated until used.

11. Anhydrous N,N-dimethylformamide was purchased from Aldrich Chemical Company, Inc.

12. tert-Butyldiphenylchlorosilane was purchased from United Chemical Technologies Inc.

13. The separation is achieved on a column of silica gel with pentane:ether (gradient 18:1 to 9:1) as the eluent.

14. The physical properties are as follows: $[\alpha]_D^{20}$ -16.4° (CHCl$_3$, c 2.8); ^{1}H NMR (300 MHz, CDCl$_3$) δ: 1.04 (s, 9 H), 1.16 (d, 3 H, J = 6.9), 2.72 (m, 1 H), 3.70 (s, 3 H), 3.73 (dd, 1 H, J = 9.9, 5.7) 3.84 (dd, 1 H, J = 9.9, 6.9), 7.36-7.46 (m, 6 H), 7.65-7.68 (m, 4 H); IR (film) cm^{-1}: 3071, 3050, 2955, 1746, 1111. Anal. Calcd for $C_{21}H_{28}O_3Si$; C, 70.74; H, 7.92. Found: C, 70.73; H, 7.80.

15. Hexane was dried by storage over activated 4 Å molecular sieves.

16. The submitters indicated that 1.1 equiv of DIBAL-H was the optimal stoichiometry to ensure complete consumption of the starting methyl ester. If less was used, the aldehyde was not readily obtained in pure form. Excess DIBAL-H results in ~5-10% overreduction to afford a small amount of the alcohol. However, the checkers found that the reaction did not go to completion under these conditions, and that it was very difficult

64

to separate the aldehyde product from the ester starting material. Therefore, the checkers used 1.13-1.15 equivalents of DIBAL-H for complete reaction, and obtained the product aldehyde in 85-86% yield along with 12-13% yields of alcohol from overreduction. The checkers also obtained small amounts of ester **1** (2-5%, depending on the batch of DIBAL-H used). Ester **1** can be visualized by TLC (R_f 0.65, 5% ether/pentane), versus Rf's = 0.40 for the starting ester and product aldehyde (which do not separate under these conditions).

17. The biphasic mixture should be stirred until both layers are clear upon settling.

18. The separation is achieved on a column of silica gel with pentane:ether (gradient 18:1 to 9:1) as the eluent. If any methyl ester remains, the aldehyde can be further purified by recrystallization from hexanes.

19. The physical properties are as follows: $[\alpha]_D^{20}$ -24.7° (CHCl$_3$, c 1.5); mp 63-64°C; ^{1}H NMR (300 MHz, CDCl$_3$) δ: 1.04 (s, 9 H), 1.10 (d, 3 H, J = 6.9), 2.57 (m, 1 H), 3.84 (dd, 1 H, J = 10.5, 6.3), 3.91 (dd, 1 H, J = 10.5, 5.4), 7.37-7.47 (m, 6 H), 7.63-7.66 (m, 4 H), 9.77 (d, 1 H, J = 1.2). Anal. Calcd for C$_{20}$H$_{26}$O$_2$Si: C, 73.57; H, 8.03. Found: C, 73.30; H, 7.93.

20. (R)-(+)-3-Butyn-2-ol was purchased from Aldrich Chemical Company, Inc., or DMS Fine Chemicals Inc.

21. Concentration of the crude mesylate under reduced pressure must be done with care to avoid loss of product due to volatility.

22. The physical properties are as follows: $[\alpha]_D^{20}$ +108.4° (CHCl$_3$, c 2.39); ^{1}H NMR (300 MHz, CDCl$_3$) δ: 1.66 (d, 3 H, J = 6.8), 2.70 (d, 1 H, J = 2.0), 3.12 (s, 3 H), 5.29 (qd, 1 H, J = 6.8, 2.0); ^{13}C NMR (75 MHz, CDCl$_3$) δ: 22.3, 39.0, 67.4, 76.6, 80.1.

23. The checkers found that the mesylate is unstable to storage and gave best results in the subsequent reaction with In(I)I if used immediately after preparation.

24. PdCl$_2$(dppf) was prepared according to the published procedure[2] from PdCl$_2$(NCPh)$_2$ and dppf ligand purchased from Aldrich Chemical Company, Inc. The commercially available PdCl$_2$(dppf) catalyst was also used, but the freshly prepared catalyst proved superior.

25. TLC analysis was performed on silica gel plates developed with hexanes:ether (3:1), R$_f$ = 0.47, and visualized with ceric(IV) sulfate/ammonium molybdate stain.

26. The separation is achieved on a column of silica gel (34 cm x 16 cm) with hexanes:ether (9:1) as the eluent.

27. The physical properties are as follows: $[\alpha]_D^{20}$-17.0° (CHCl$_3$, c 1.51); ^{1}H NMR (300 MHz, CDCl$_3$) δ: 0.86 (d, 3 H, J = 6.9), 1.07 (s, 9 H), 1.33 (d, 3 H, J = 7.5), 2.06 (m, 1 H), 2.15 (d, 1 H, J = 2.4), 2.71 (m, 1 H), 3.42 (m, 1 H), 3.44 (d, 1 H, J = 1.5), 3.71 (dd, 1 H, J = 10.2, 6.9), 3.78 (dd, 1 H, J = 10.2, 4.2), 7.37-7.46 (m, 6 H), 7.72-7.67 (m, 4 H); IR (film) cm^{-1}: 3493, 3305, 2931; ^{13}C NMR (75 MHz, CDCl$_3$) δ: 13.5, 17.9, 19.1, 26.8, 30.2, 38.9, 68.6, 70.2, 78.1, 85.0, 127.8, 129.8, 132.9, 135.6. Anal. Calcd for C$_{24}$H$_{32}$O$_2$Si: C, 75.74; H, 8.47. Found: C, 75.74; H, 8.43.

28. The (2R,3R,4R) diastereomer results from partial racemization of one or both of the allenylmetal intermediates. This point was confirmed by comparison to authentic material as the (S)-MPA ((S)-(2-methoxy)phenylacetic acid-Mosher's acid) derivative. The optical rotations of these compounds are small, and thus correlation by comparison of $[\alpha]_D^{20}$ values is unreliable.

Waste Disposal Information

All toxic materials were disposed of in accordance with "Prudent Practices in the Laboratory"; National Academy Press; Washington, DC, 1995.

3. Discussion

Chiral allenylmetal compounds provide convenient access to enantioenriched homopropargylic alcohols through S_E2' additions to aldehydes.[3] The syn adducts can be obtained through addition of allenyl tributylstannanes in the presence of stoichiometric boron trifluoride etherate ($BF_3 \cdot OEt_2$). The use of allenylmetal halides derivatives of Sn, Zn, and In lead to the anti diastereomers. The former additions proceed through an acyclic transition state whereas the latter are thought to involve a cyclic transition state, thus accounting for the difference in diastereoselectivity.

The present method is practical and efficient as it employs readily available enantioenriched propargylic alcohols[4] as precursors to the allenylindium reagents. With achiral aldehydes the diastereoselectivity is high for branched aldehydes, moderate for unbranched aldehydes, and low for benzaldehyde (Table I).[5] With chiral α-methyl aldehydes[6] the additions proceed under effective reagent control to afford anti adducts of high ee and with excellent diastereoselectivity (eq. 1 and 2). Comparable results were obtained with 3:1 dimethyl sulfoxide-tetrahydrofuran (DMSO-THF) as the solvent.

Table I

Additions of Transient Chiral Allenylindium Reagents To Representative Achiral Aldehydes[5]

R	yield, %	anti:syn [b]	ee, %[c]
c-C_6H_{11}	79	95:5	89
i-Pr	74	94:6	95
C_6H_{13}	77	84:16	95
C_6H_5	85	45:55	---

[a] dppf = diphenylphosphinoferrocene; [b] analysis by gas chromatography; [c] major isomer

$$\text{(1)}$$

R^1 = H, R^2 = DPS; 87%, >95:5 anti,anti:anti,syn
R^1 = H, R^2 = TBS; 78%, >95:5 anti,anti:anti,syn *
R^1 = H, R^2 = Bn; 96%, 90:10 anti,anti:anti,syn *
R^1 = H, R^2 = PMB; 84%, 80:20 anti,anti:anti,syn
R^1 = CH$_2$OBn, R^2 = DPS; 83%, >95:5 anti,anti:anti,syn

* The enantiomeric aldehyde and mesylate were employed. The products are enantiomeric to those depicted.

$$\text{(2)}$$

R^1 = H, R^2 = DPS; 88%, >95:5 anti,syn:anti,anti
R^1 = H, R^2 = PMB; 70%, 83:17 anti,syn:anti,anti
R^1 = CH$_2$OBn, R^2 = DPS; 87%, >95:5 anti,syn:anti,anti

The described preparation of InI is a modification of a 3-step literature procedure in which indium shot is hammered into indium foil and heated with iodine to form InI$_3$.[7] The InI$_3$ is then heated with excess indium foil to form In$_2$I$_4$ (as a complex of InI[InI$_3$]).In a third step, the In$_2$I$_4$ is treated with diethyl ether whereupon it disproportionates to insoluble InI and soluble InI$_3$ etherate that are separated by filtration.

The present procedure combines the second and third steps, thus avoiding handling of the hygroscopic intermediate In$_2$I$_4$ complex. It also demonstrates efficient recycling of the recovered InI$_3$ etherate. Commercial InI, available from Aldrich Chemical Company, Inc., can be used in the allenylindium procedure with comparable results but at greater expense. The InI prepared as described is an easily handled free-flowing powder, whereas the commercial product consists of small beads that must be crushed before use.

An alternative preparation of enantioenriched anti-homopropargylic alcohols along similar lines uses Et_2Zn to effect in situ transmetallation of an allenylpalladium intermediate from a propargyl mesylate and 2.5 mol % of $Pd(OAc)_2 \cdot PPh_3$ in the presence of an aldehyde (Table II).[8] The two methods are comparable.

Table II.
Additions of Transient Allenylzinc Reagents to Representative Achiral Aldehydes[8]

R	yield, %	anti:syn [a]	ee, % [b]
c-C_6H_{11}	75	95:5	90
i-Pr	75	95:5	92
C_6H_{13}	74	88:12	95

[a] analysis by gas chromatography; [b] major isomer

1. Department of Chemistry, University of Virginia, Charlottesville, VA 22901.

2. (a) Heck, R. F. "Palladium Reagents in Organic Synthesis"; Academic Press: London, 1985; p. 17; (b) Hiyashi T.; Konishi, M.; Kobori, Y.; Kumada, M.; Higuchi, T.; Hirotsu, K. J. Am. Chem. Soc. 1984, 106, 158-163.

3. Marshall, J. A. Chem. Rev. 1996, 96, 31.

4. These can be prepared by a) asymmetric dihydroxylation of allylic chlorides followed by acetonide formation and elimination with BuLi; b) catalytic asymmetric transfer hydrogenation and c) reduction of alkynones with chiral metal hydrides. (a) Marshall, J. A.; Jiang, H. Tetrahedron Lett. 1998, 39, 1493; Yadav, J. S.; Chander, M. C.; Rao, C. S. Tetrahedron Lett. 1989, 30, 5455; (b) Matsumura, K.; Hashiguchi, S.; Ikariya, T.; Noyori, R. J. Am. Chem. Soc. 1997, 119, 8738; (c) Marshall, J. A.; Wang, X. -j. J. Org. Chem. 1991, 56, 3211.

5. Marshall, J. A.; Grant, C. M. J. Org. Chem. 1999, 64, 696.

6. Heathcock, C. H.; Young, S. D.; Hagen, J. P.; Pilli, R.; Badertscher, U. J. Org. Chem. 1985, 50, 2095; Wasicak, J. T.; Donaldson, W. A. Tetrahedron: Asymmetry

1998, *9*, 133; Roush, W. R.; Palkowitz, A. D.; Palmer, M. A. J. *J. Org. Chem.* **1987**, *52*, 316.

7.   This procedure is a modification of the method of Tuck: Freeland, B. H.; Tuck, D. G. *Inorg. Chem.* **1976**, *15*, 475.

8.   Marshall, J. A.; Adams, N. D. *J. Org. Chem.* **1998**, *63*, 3812.

Appendix
Chemical Abstracts Nomenclature (Collective Index Number);
(Registry Number)

Indium (I) iodide: Indium iodide (8); Indium iodide (InI) (9); (13966-94-4)

(2R,3S,4S)-1-(tert-Butyldiphenylsilyloxy)-2,4-dimethyl-5-hexyn-3-ol: 5-Hexyn-3-ol, 1-[[(1,1-dimethylethyl)diphenylsilyl]oxy]-2,4-dimethyl-, (2R,3S,4S)- (14); (220634-80-0)

Indium (III) iodide: Indium iodide (8); Indium iodide (InI$_3$) (9); (13510-35-5)

Indium (8, 9); (7440-74-6)

Iodine (8, 9); (7553-56-2)

(R)-3-(tert-Butyldiphenylsilyloxy)-2-methylpropanal: Propanal, 3-[[(1,1-dimethylethyl)diphenylsilyl]oxy]-2-methyl-, (2R)- (12); (112897-04-8)

Methyl (R)-(-)-3-hydroxy-2-methylpropionate: Propanoic acid, 3-hydroxy-2-methyl-, methyl ester, (R)- (10); (72657-23-9)

N,N-Dimethylformanide: CANCER SUSPECT AGENT: Formanide, N,N-dimethyl- (8, 9); (68-12-2)

Imidazole (8); 1H-Imidazole (9); (288-32-4)

tert-Butyldiphenylchlorosilane: Silane, chloro(1,1-dimethylethyl)diphenyl- (9); (58479-61-1)

Methyl (R)-3-(tert-butyldiphenylsilyloxy)-2-methylpropionate: Propanoic acid, [[(1,1-dimethylethyl)diphenylsilyl]oxy]-2-methyl-, methyl ester, (2R)- (13); (153775-90-7)

Diisobutylaluminum hydride: DIBAL-H: Aluminum, hydrodiisobutyl- (8); Aluminum, hydrobis(2-methylpropyl)- (9); (1191-15-7)

(R)-3-Butyn-2-yl methanesulfonate: 3-Butyn-2-ol, methanesulfonate, (2R)- (12); (121887-95-4)

(R)-(+)-3-Butyn-2-ol: 3-Butyn-2-ol, (+)- (9); (42969-65-3)

Triethylamine (8); Ethanamine, N,N-diethyl- (9); (121-44-8)

Methanesulfonyl chloride (8, 9); (124-63-0)

Hexamethylphosphoramide: HIGHLY TOXIC; CANCER SUSPECT AGENT: Phosphoric triamide, hexamethyl- (8, 9); (680-31-9)

[1,1'-Bis(diphenylphosphino)ferrocene]dichloropalladium: Palladium, [1,1'-bis(diphenylphosphino)ferrocene-P,P']dichloro- (10); (72287-26-4)

Bis(benzonitrile)dichloropalladium(II): Palladium, bis(benzonitrile)dichloro- (8, 9); (14220-64-5)

1,1'-Bis(diphenylphosphino)ferrocene (dppf): Phosphine, 1,1'-ferrocenediylbis[diphenyl- (8); Ferrocene, 1,1'-bis(diphenylphosphino)- (9); (12150-46-8)

(2R,3R,4R)-1-(tert-Butyldiphenylsilyloxy)-2,4-dimethyl-5-hexyn-3-ol: 5-Hexyn-3-ol, 1-[[(1,1-dimethylethyl)diphenylsilyl]oxy]-2,4-dimethyl-, (2R,3R,4R)- (14); (220634-81-1)

ASYMMETRIC SYNTHESIS OF (M)-2-HYDROXYMETHYL-1-(2-HYDROXY-4,6-DIMETHYLPHENYL)NAPHTHALENE VIA A CONFIGURATIONALLY UNSTABLE BIARYL LACTONE

[2-Naphthalenemethanol, 1-(2-hydroxy-4,6-dimethylphenyl)-, (R)-]

A.

B.

C.

Submitted by Gerhard Bringmann,[1] Matthias Breuning, Petra Henschel, and Jürgen Hinrichs.

Checked by Peter M. Greenen and Dennis P. Curran

1. Procedure

A. *3,5-Dimethylphenyl 1-bromo-2-naphthoate (3)*. Under a nitrogen atmosphere, a 250-mL, oven-dried, round-bottomed flask containing anhydrous dichloromethane (100 mL) is charged with 1-bromo-2-naphthoic acid (1, 2.51 g, 10.0 mmol), 3,5-dimethylphenol (2, 1.23 g, 10.1 mmol), dicyclohexylcarbodiimide (DCC, 2.26 g, 11.0 mmol), and 4-(dimethylamino)pyridine (DMAP, 244 mg, 2.00 mmol) (Note 1). After the mixture is stirred for 12 hr at room temperature, the white precipitate that forms (Note 2) is discarded by filtration through a Buchner funnel. From the clear filtrate, the solvent is removed by rotary evaporation (35°C, 720 mbar, 540 mm) to give a colorless solid. Filtration through a short silica gel column (5 x 40-cm column, silica gel 0.063 - 0.2 mm, 150 g; eluent: hexane / diethyl ether 5:1) delivers 3.35 g (94%) of the ester 3, which is recrystallized from diethyl ether / hexane to give 3.28 g (92%) of a colorless solid (Note 3).

B. *1,3-Dimethyl-6H-benzo[b]naphtho[1,2-d]pyran-6-one (4)*. Under an argon atmosphere, a 250-mL, oven-dried, round-bottomed flask, equipped with a reflux condenser, is charged with freshly distilled N,N-dimethylacetamide (DMA, 130 mL), 3,5-dimethylphenyl 1-bromo-2-naphthoate (3, 3.24 g, 9.12 mmol), palladium(II) acetate (205 mg, 0.913 mmol), triphenylphosphine (481 mg, 1.83 mmol), and sodium acetate (1.50 g, 18.3 mmol) (Notes 4 and 5). The orange suspension is degassed three times, placed in a preheated (130°C) oil bath (Note 5), and stirred at 130°C for 12 hr (Note 6). Removal of the solvent at 40°C (0.1 mbar, 0.075 mm) gives a black oily residue, which is chromatographed (5 x 40 cm column, silica gel 0.063 - 0.2 mm, 170 g, 1 cm of charcoal at the top of the column; eluent: hexane / diethyl ether 5:1), to yield 2.00 g (80%) of the lactone 4 as a slightly yellow solid. Recrystallization from diethyl ether / hexane delivers 1.63 g (65%) of colorless or pale yellow crystals (Note 7).

C. (M)-2-Hydroxymethyl-1-(2-hydroxy-4,6-dimethylphenyl)naphthalene   [(M)-6]
(Note 8). Under an argon atmosphere, an oven-dried Schlenk tube is charged with the CBS-catalyst (S)-**5** (1.0 M in toluene, 8.39 mL, 8.39 mmol) (Note 9). The solvent is removed under high vacuum (0.1 mbar. 0.075 mm) at room temperature and tetrahydrofuran (THF, 110 mL) (Note 9) is added. After the solution is cooled to 0°C, it is treated with the borane-THF complex (1.0 M in THF, 10.1 mL, 10.1 mmol) (Note 9) and stirred at room temperature for 30 min. This reagent and a solution of the lactone **4** (1.84 g, 6.71 mmol) in THF (110 mL) are added simultaneously from two dropping funnels into an oven-dried, round-bottomed, three-necked flask containing THF (110 mL) at a temperature of 30°C over a period of 2 hr (Note 10). After the reaction mixture is stirred for another 30 min, it is adjusted to pH 4 by careful addition of hydrochloric acid (2.0 M). Water (20 mL) is added, the organic solvent is removed by rotary evaporation (40°C / 350 mbar, 263 mm), and the remaining aqueous phase is extracted with diethyl ether (4 x 100 mL) (Note 11). The combined organic layers are dried over magnesium sulfate ($MgSO_4$). Removal of the solvent by rotary evaporation and filtration through a short silica gel column (5 x 40-cm column, silica gel 0.063 - 0.2 mm, 100 g; eluent: hexane / diethyl ether 1:1) gives 1.83 g (98%) of the biaryl alcohol (M)-**6** as a slightly yellow solid with 92% ee (Notes 12 and 13). Crystallization from diethyl ether / hexane delivers 1.53 g (82%) of colorless crystals with >99% ee (Notes 14 and 15).

2. Notes

1. 1-Bromo-2-naphthoic acid (**1**, 98%, Sigma Chemical Co.), 3,5-dimethylphenol (**2**, >98%, Merck-Schuchard), 1,3-dicyclohexylcarbodiimide (DCC, 99%, Merck-Schuchard), and 4-(dimethylamino)pyridine (DMAP, 99%, Aldrich Chemical Co., Inc.) were used as received. Dichloromethane was distilled from phosphorus pentoxide and stored over activated molecular sieves (4Å).

74

2. The precipitate consists of 1,3-dicyclohexylurea.

3. The physical properties of **3** are as follows: mp 85°C; IR (KBr) cm^{-1}: 1594, 1618, 1744, 2916, 3060; ^{1}H NMR (250 MHz, CDCl$_3$) δ: 2.38 (s, 6 H), 6.95 (s, 3 H), 7.67 (m$_c$, 2 H), 7.88 (m$_c$, 3 H), 8.50 (m$_c$, 1 H); ^{13}C NMR (63 MHz, CDCl$_3$) δ: 21.30, 119.1, 123.3, 125.9, 127.9, 128.0, 128.0, 128.2, 128.4, 128.7, 130.7, 132.6, 135.4, 139.5, 150.7, 166.0; HRMS: m/z 354.0257, calcd. for C$_{19}$H$_{15}$BrO$_2$: 354.0255.

4. Palladium(II) acetate (98%, Strem Chemicals Inc.), triphenylphosphine (99%, Fisher Scientific Co.), and sodium acetate (99%, Fluka Chemika) were used without further purification. N,N-Dimethylacetamide (DMA) was distilled from calcium hydride through a 25-cm Vigreux column directly before use.

5. Freshly distilled DMA, a preheated oil bath, and the repeated degassing of the reaction mixture are critical to obtain high yields.

6. The reaction progress can easily be followed by TLC (silica, hexane / diethyl ether 5:1) due to the brilliant blue fluorescence of **4** (R$_f$ = 0.30) upon UV excitation at 366 nm.

7. The physical properties of **4** are as follows: mp 158°C; IR (KBr) cm^{-1}: 1594, 1614, 1721, 2928, 2985, 3055; ^{1}H NMR (250 MHz, CDCl$_3$) δ: 2.25 (s, 3 H), 2.46 (s, 3 H), 7.08 (s, 1 H), 7.15 (s, 1 H), 7.56 (m, 1 H), 7.66 (m, 1 H), 7.96 (d, 3 H, J = 8.5 Hz), 8.27 (d, 1 H, J = 8.5 Hz); ^{13}C NMR (63 MHz, CDCl$_3$) δ: 21.27, 23.77, 114.7, 116.0, 121.2, 124.0, 125.9, 128.1, 128.3, 128.6, 128.8, 128.9, 135.6, 136.3, 136.3, 140.1, 140.2, 151.9, 161.9; HRMS: m/z 274.0987, calcd. for C$_{19}$H$_{14}$O$_2$: 274.0994.

8. For the now recommended M/P denotion for axial chirality, see Helmchen[2].

9. The CBS-catalyst [(S)-2-methyl-CBS-oxazaborolidine] (S)-**5** (1.0 M in toluene) (The CBS catalyst is named after Corey, Bakshi, and Shibata) and the borane-THF complex (1.0 M in THF) were obtained from Aldrich Chemical Co., Inc. and used as received. THF was distilled from potassium directly before use.

10. For optimum stereoselectivity, it is critical to control the temperature of the reaction vessel to exactly 30°C.

11. For a recovery of (S)-α,α-diphenylprolinol, which is the hydrolysis product of the CBS-catalyst (S)-5 (and likewise its synthetic precursor[3]), the aqueous phase is carefully adjusted to pH 10 with concentrated ammonia and extracted with diethyl ether (3 x 50 mL). The combined organic layers are washed with brine (50 mL) and dried over $MgSO_4$. Removal of the solvent by rotary evaporation yields 1.68 g (79%) of crude (S)-α,α-diphenylprolinol. This material is dissolved in dichloromethane / methanol 9:1 (3 mL) and filtered over Alox B (act. III, 80 g) with dichloromethane / methanol 9:1 as the eluent, to yield 1.64 g (77%) of (S)-α,α-diphenylprolinol as a white solid.

12. The ee of 6 was determined by HPLC on a chiral phase [DAICEL Chiralcel OD-H (4.6 mm x 250 mm), detection at 280 nm, flow rate 1.0 mL/min, eluent: hexane / isopropyl alcohol 95:5, retention times: t_R = 16 min for (M)-6 and t_R = 22 min for (P)-6].

13. The analogous reduction of 4 with a catalytic amount of the CBS-catalyst (S)-5 (0.1 equiv) and 1.25 equiv of borane resulted in the formation of (M)-6 in a slightly lower ee of 88% (94% yield). In several cases, over-stoichiometric amounts of (S)-5 and $BH_3 \cdot$ THF had to be used to ensure complete conversion.[4-6] Treatment of 4 with 3 equiv of (S)-5 and 4 equiv of the borane complex gave 97% of (M)-6 with 90% ee.

14. The almost racemic alcohol 6 obtained from the concentrated mother liquor can be recycled by a three-step procedure (1. MnO_2, CH_2Cl_2; 2. $NaClO_2$, H_2NSO_3H, NaOAc, dioxane / acetic acid / water; 3. N-methyl-2-chloropyridinium iodide, (n-Bu)$_3$N, CH_2Cl_2) to give, in a 49% overall yield, the lactone 4, which can then be ring-opened atropoenantioselectively once again. Alternatively, the recycling can be done by just oxidizing to the corresponding hydroxy aldehyde, followed by its atropo-enantioselectively reduction.[4,7]

15. The physical properties of 6 are as follows: mp 140°C; $[\alpha]_D^{23}$ -41.4° ($CHCl_3$, c 1.02); IR (KBr) cm^{-1}: 1572, 1620, 2923, 2986, 3055, 3376; ^{1}H NMR (200 MHz, $CDCl_3$)

δ: 1.80 (s, 3 H), 2.38 (s, 3 H), 4.52 (m_c, 2 H), 6.73 (s, 1 H), 6.79 (s, 1 H), 7.32-7.55 (m, 3 H), 7.70 (d, 1 H, J = 8.5), 7.92 (m_c, 2 H); ^{13}C NMR (50 MHz, $CDCl_3$) δ: 19.73, 21.28, 63.70, 114.0, 121.0, 123.4, 125.4, 126.2, 126.5, 126.8, 128.2, 128.9, 131.2, 132.5, 133.5, 137.4, 138.0, 139.2, 153.1; HRMS: m/z 278.1304, calcd. for $C_{19}H_{18}O_2$: 278.1307.

Waste Disposal Information

All toxic materials were disposed of in accordance with "Prudent Practices in the Laboratory"; National Academy Press; Washington, DC, 1995.

3. Discussion

The reaction sequence described here provides a simple and efficient route to the enantiomerically pure axially chiral biaryl alcohol (M)-**6**[8] and illustrates the basic strategy of the 'lactone concept',[9] in which the two crucial steps, the aryl-aryl bond formation and asymmetric induction at the newly created axis, are performed *consecutively*. This stepwise procedure is quite generally applicable and offers several advantages over other known methods[10] of stereoselective biaryl coupling: The prefixation of the aryl moieties as esters of type **3** (Scheme 1), which is easily attainable by standard procedures, allows an intramolecular cross coupling. This coupling reaction proceeds regioselectively and in high yields, even against severe steric hindrance (e.g., with a tert-butyl group ortho to the axis).[11,12] As the catalyst, Pd(OAc)$_2$ or the more effective, now likewise commercially available Herrmann-Beller palladacycle **7** can be employed.[11,12] The resulting biaryl lactones **4** are helically distorted and thus chiral,[11] but because of the bridging lactone function, which dramatically lowers the atropoisomerization barrier, they are still configurationally unstable at the axis (An exception is the sterically highly hindered lactone **4** (R = tBu), which is configurationally stable at room temperature.), and thus exist as a racemic mixture of their rapidly interconverting enantiomers (M)-4 and (P)-4.[11,13] This is the fundamental prerequisite for the subsequent formation of configurationally stable and stereochemically pure biaryl molecules 6 through oxazaborolidine-mediated dynamic kinetic resolution[8,12] (see Scheme 1).

Scheme 1

o = configurationally unstable
(except for R = tBu)

cat.:

Ar = o-tolyl   7

	coupling[8,10]	ring cleavage[7,11]	
entry	R	yield [%]	yield [%]	ee [%]
1	H	70 (91)[a]	94	82
2	OMe	77 (90)	95	94[b]
3	Me	74 (90)	98	92
4	tBu	31 (81)	c	99[c]

[a] method (B) in parantheses
[b] note that for formal reasons of the CIP denotation, stereochemically analogous biaryls of this series with R = OMe will have descriptors opposite
to those with R = H or alkyl
[c] at the beginning of the reaction (kinetic resolution, $k_{rel} > 200$)[11]

In the stereochemically deciding key step, cleavage of the lactone bridge can also be performed atropo-diastereo- or -enantioselectively with a wide range of chiral O-,[14] N-,[15] or other H-[16] nucleophiles to give the ring-opened, and now configurationally stable, axially chiral biaryls in high optical and chemical yields. In each case, the stereochemically pure biaryls can be obtained by crystallization or, if diastereomers are formed, by chromatographic separation. Two examples are illustrated in Scheme 2 with 4 (R = Me) as the biaryl lactone.

Scheme 2

Since, for all the chiral ring cleavage reagents [(S)-**5**, (R)-**10**, and (S)-**11**] used, both enantiomers are commercially available, both atropoisomeric biaryls are readily accessible from the same lactone precursor **4**, which allows flexible atropo-divergent[9] syntheses.[14,15] Furthermore, for precious material prepared in the course of multi-step synthesis, even the minor, undesired atropoisomer, if formed at all in significant quantities, is not lost, but can be recycled, either by acid catalyzed- (for the ester **8** or the amide **9**) or oxidative (for the alcohol **6**) cyclization back to the biaryl lactone **4**, and renewed atroposelective cleavage.[14,15] Differing from most of the existing methods, the decision as to which atropoisomer is to be prepared can be taken at a very late stage of the synthesis.

The lactone concept is not restricted to the simple model biaryl synthesis presented here. It has been successfully expanded to a broad series of structurally diverse biaryl substrates (e.g., lactones with additional stereocenters and functional groups,[9] configurationally stable lactones,[12] seven-membered lactones,[17] and again configurationally unstable biaryl hydroxy aldehydes[7]), to different activation modes in the ring-opening step (e.g., use of metallated nucleophiles, carbonyl activation by Lewis acids, (η^6-complexation, etc.),[18] and for various strategies of stereoselection (e.g., external vs. internal asymmetric induction).[19]

The broad applicability of the strategy has been proven in the atroposelective synthesis of a broad series of structurally different bioactive natural biaryl products like the naphthylisoquinoline alkaloid dioncopeltine A (**16**)[4] (Scheme 3), the dimeric sesquiterpene mastigophorene A,[6] the phenyl anthraquinone knipholone,[20] or even molecules without O- or (free) C_1-units next to the axis, i.e., dioncophylline C[19] and korupensamine A.[5] Furthermore, biaryl derivatives resulting from the ring opening of the model lactone **4** were used successfully as chiral catalysts in asymmetric synthesis, like amino alcohols for the enantioselective addition of Et_2Zn to aldehydes[21] or a phosphine for the enantioselective hydrosilylation of styrenes.[22]

Scheme 3

12 R = *i*Pr, R' = Bn

BCl₃ ⎡ **13** R = *i*Pr, R' = Bn
     ⎣ **14** R = H, R' = Bn

H₂, Pd/C ⎡ (*M*)-**15** R' = Bn
         ⎣ (*M*)-**16** R' = H

1. Institut für Organische Chemie, Universität Würzburg, Am Hubland, D-97074 Würzburg, Germany.

2. Helmchen, G. In "Methods of Organic Chemistry (Houben Weyl)", 4th ed.; Vol. E21a; Helmchen, G.; Hoffmann, R. W.; Mulzer, J.; Schaumann, E., Ed.; Thieme: Stuttgart, 1995; p 11.

3. (a) Xavier, L. C.; Mohan, J. J.; Mathre, D. J.; Thompson, A. S.; Carroll, J. D.; Corley, E. G.; Desmond, R. *Org. Synth.* **1997**, *74*, 50; (b) Corey, E. J.; Helal, C. J. *Angew. Chem., Int. Ed. Engl.* **1998**, *37*, 1986.

4. Bringmann, G.; Saeb, W.; Rubenacker, M. *Tetrahedron* **1999**, *55*, 423.

5. Bringmann, G.; Ochse, M.; Götz, R. *J. Org. Chem.* **2000**, *65*, 2069.

6. Bringmann, G.; Pabst, T.; Henschel, P.; Kraus, J.; Peters, K.; Peters, E.-M.; Rycroft, D. S.; Connolly, J. D. *J. Am. Chem. Soc.,* submitted.

7. Bringmann, G.; Breuning, M. *Synlett* **1998**, 634.

8. (a) Bringmann, G.; Hartung; T. *Angew. Chem., Int. Ed. Engl.* **1992**, *31*, 761; (b) Bringmann, G.; Hartung, T. *Tetrahedron* **1993**, *49*, 7891.

9. (a) Bringmann, G.; Breuning, M.; Tasler, S. *Synthesis* **1999**, 525; (b) Bringmann, G.; Tasler, S. In "Current Trends in Organic Synthesis"; Scolastico, C.; Nicotra, F., Eds.; Plenum Publishing Corporation: New York, 1999, p. 105.

10. (a) Gant, T. G.; Meyers, A. I. *Tetrahedron* **1994**, *50*, 2297; (b) Feldman, K. S.; Smith, R. S. *J. Org. Chem.* **1996**, *61*, 2606; (c) Kamikawa, K.; Watanabe, T.; Uemura, M. *J. Org. Chem.* **1996**, *61*, 1375; (d) Lipshutz, B. H.; Kayser, F.; Liu, Z.-P. *Angew. Chem., Int. Ed. Engl.* **1994**, *33*, 1842; (e) Miyano, S.; Fukushima, H.; Handa, S.; Ito, H.; Hashimoto, H. *Bull. Chem. Soc. Jpn.* **1988**, *61*, 3249.

11. Bringmann, G.; Hartung, T.; Göbel, L.; Schupp, O.; Ewers, C. L. J.; Schöner, B.; Zagst, R.; Peters, K.; von Schnering, H. G.; Burschka, C. *Liebigs Ann. Chem.* **1992**, 225.

12. Bringmann, G.; Hinrichs, J.; Kraus, J.; Wuzik, A.; Schulz, T. *J. Org. Chem.* **2000**, *65*, 2517.

13. Bringmann, G.; Heubes, M.; Breuning, M.; Göbel, L.; Ochse, M.; Schöner, B.; Schupp, O. *J. Org. Chem.* **2000**, *65*, 722.

14. (a) Bringmann, G.; Breuning, M.; Walter, R.; Wuzik, A.; Peters, K.; Peters, E.-M. *Eur. J. Org. Chem.* **1999**, 3047; (b) Seebach, D.; Jaeschke, G.; Gottwald, K.; Matsuda, K.; Formisano, R.; Chaplin, D. A.; Breuning, M.; Bringmann, G. *Tetrahedron* **1997**, *53*, 7539.

15. Bringmann, G.; Breuning, M.; Tasler, S.; Endress, H.; Ewers, C. L. J.; Gobel, L.; Peters, K.; Peters, E.-M. *Chem. Eur. J.* **1999**, *5*, 3029.

16. Bringmann, G.; Breuning, M. *Tetrahedron: Asymmetry* **1999**, *10*, 385.

17. Bringmann, G., Hinrichs, J. *Tetrahedron: Asymmetry* **1997**, *8*, 4121.

18. (a) Schenk, W. A.; Kümmel, J.; Reuther, I.; Burzlaff, N.; Wuzik, A.; Schupp, O.; Bringmann, G. *Eur. J. Inorg. Chem.* **1999**, 1745; (b) Bringmann, G.; Wuzik, A.; Stowasser, R.; Rummey, C.; Göbel, L.; Stalke, D.; Pfeiffer, M.; Schenk, W. A. *Organometallics* **1999**, *18*, 5017.

19. Bringmann, G.; Holenz, J.; Weirich, R.; Rubenacker, M.; Funke, C.; Boyd, M. R.; Gulakowski, R. J.; Francois, G. *Tetrahedron* **1998**, *54*, 497.

20. Bringmann, G.; Menche, D., manuscript in preparation.

21. Bringmann, G.; Breuning, M. *Tetrahedron: Asymmetry* **1998**, *9*, 667.

22. Bringmann, G.; Wuzik, A.; Breuning, M.; Henschel, P.; Peters, K.; Peters E.-M. *Tetrahedron: Asymmetry* **1999**, *10*, 3025.

Appendix
Chemical Abstracts Nomenclature (Collective Index Number);
(Registry Number)

(M)-2-Hydroxymethyl-1-(2-hydroxy-4,6-dimethylphenyl)naphthalene:

2-Naphthalenemethanol, 1-(2-hydroxy-4,6-dimethylphenyl)-, (R)- (13); (140834-52-2)

3,5-Dimethylphenyl-1-bromo-2-naphthoate: 2-Naphthalenecarboxylic acid,

1-bromo-, 3,5-dimethylphenyl ester (13); (138435-66-2)

1-Bromo-2-naphthoic acid: 2-Naphthoic acid, 1-bromo- (9); (20717-79-7)

3,5-Dimethylphenol: Phenol, 3,5-dimethyl- (9); (108-68-9)

Dicyclohexylcarbodiimide: HIGHLY TOXIC: Carbodiimide, dicyclohexyl- (8);

Cyclohexanamine, N,N'-methanetetraylbis- (9); (538-75-0)

4-Dimethylaminopyridine: HIGHLY TOXIC: Pyridine, 4-(dimethylamino)- (8); 4-Pyridinamine, N,N-dimethyl- (9); (1122-58-3)

1,3-Dimethyl-6H-benzo[b]naphtho[1,2-d]pyran-6-one: 6H-Benzo[b]naphtho[1,2-d]pyran-6-one, 1,3-dimethyl- (13); (138435-72-0)

N,N-Dimethylacetamide: Acetamide, N,N-dimethyl- (8,9); (127-19-5)

Palladium acetate: Acetic acid, palladium(2+) salt (8,9); (3375-31-3)

Triphenylphosphine: Phosphine, triphenyl- (8,9); (603-35-0)

(S)-2-Methyl-CBS-oxazaborolidine: (CBS named after Corey, Bakshi, Shibata): 1H, 3H-Pyrrolo[1,2-c][1,3,2]oxazaborole, tetrahydro-1-methyl-3,3-diphenyl-, (S)- (12); (112022-81-8)

Borane-tetrahydrofuran complex: Furan, tetrahydro-, compd. with borane (1:1) (8,9); (14044-65-6)

(R)-3-PHENYLCYCLOHEXANONE

[Cyclohexanone, 3-phenyl-, (R)-]

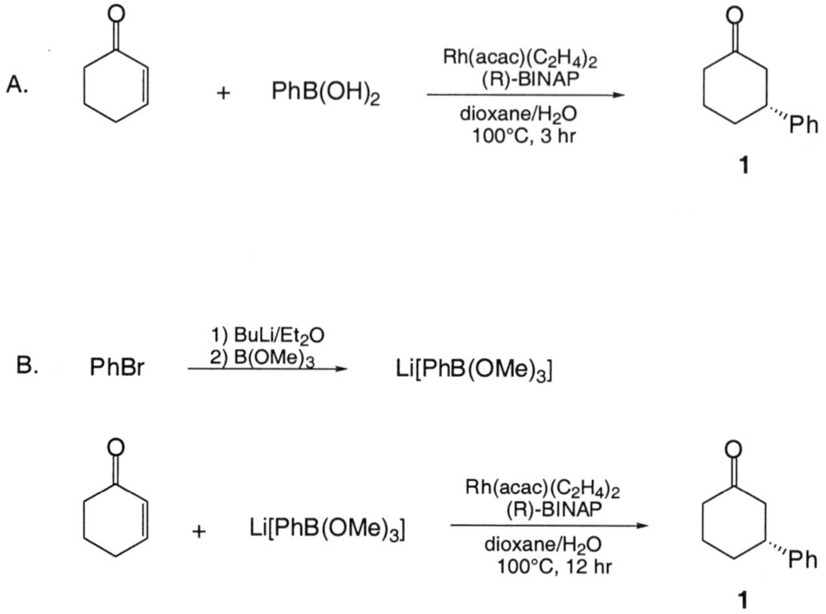

A.

$$Rh(acac)(C_2H_4)_2$$
$$(R)\text{-BINAP}$$

dioxane/H_2O
100°C, 3 hr

1

B. PhBr

1) BuLi/Et_2O
2) $B(OMe)_3$

$Li[PhB(OMe)_3]$

+ $Li[PhB(OMe)_3]$

$$Rh(acac)(C_2H_4)_2$$
$$(R)\text{-BINAP}$$

dioxane/H_2O
100°C, 12 hr

1

Submitted by Tamio Hayashi, Makoto Takahashi, Yoshiaki Takaya, and Masamichi Ogasawara.[1]

Checked by Timothy B. Durham and Marvin J. Miller.

1. Procedure

Caution! All reactions should be conducted in a well-ventilated hood.

(R)-(+)-3-Phenylcyclohexanone (**1**).

Method A. A 500-mL, two-necked, round-bottomed flask, fitted with a magnetic stirring bar, rubber septum, and a reflux condenser attached to a gas-flow adapter with a stopcock, is charged with 12.2 g (100 mmol) of phenylboronic acid (Notes 1, 2), 300 mg (0.482 mmol) of (R)-2,2'-bis(diphenylphosphino)-1,1'-binaphthyl [(R)-BINAP] (Notes 3, 4), and 103 mg (0.399 mmol) of acetylacetonatobis(ethylene)rhodium(I) (Note 5), and the flask is flushed with nitrogen. To the flask are added 3.86 g (3.90 mL, 40.2 mmol) of 2-cyclohexenone (Note 6), 200 mL of 1,4-dioxane (Note 7), and 20 mL of water via syringe, and the entire orange mixture is immersed in a oil bath preheated to 120°C and heated at 105°C for 3 hr. After the solvent is cooled to room temperature, it is removed under reduced pressure on a rotary evaporator. After concentration of the initial reaction mixture on a rotary evaporator, the residue is dissolved in diethyl ether (100 mL). The resulting solution is transferred to a 500-mL separatory funnel and washed with 1.2 M hydrochloric acid (HCl, 100 mL) followed by 5% sodium hydroxide (NaOH) solution (100 mL). The aqueous washes are separately extracted with diethyl ether (30 mL each). The ether layers are combined and washed with saturated sodium chloride solution (100 mL), dried over anhydrous magnesium sulfate, filtered, and the filtrate concentrated on a rotary evaporator to give a brown oil. The oil is filtered through silica (200 mL) with hexanes (500 mL) followed by diethyl ether (400 mL) on a 4-cm φ column. Fractions of eluent (20 mL) are collected and those containing the product as determined by TLC are combined and concentrated on a rotary evaporator to give a brown oil that is distilled under reduced pressure (0.5 mm) to give 5.77 g (83% yield) of (R)-3-phenylcyclohexanone (1) (Note 8) as a colorless oil. The enantiomeric purity is 98.6% ee, which is determined by HPLC analysis using a chiral stationary phase column (Note 9).

Method B. A dry, 200-mL, two-necked, round-bottomed flask fitted with a magnetic stirring bar, rubber septum, and a reflux condenser attached to a gas-flow adapter with a stopcock, is flushed with nitrogen. The flask is charged with 3.95 g (25.2 mmol) of bromobenzene (Note 10) and 12.5 mL of dry diethyl ether (Note 11) via syringe, and cooled to 0°C in an ice-water bath. At this temperature, 16.7 mL (25.1 mmol) of a 1.50 M solution of n-butyllithium in hexane (Note 12) is added. The ice-water bath is removed and the mixture is stirred at room temperature for 1 hr. The flask is cooled to −78°C in a dry ice-acetone bath, and 2.59 g (24.9 mmol) of trimethoxyborane (Note 13) is added dropwise over a period of 10 min via syringe. The mixture is stirred at −78°C for 30 min and slowly warmed to room temperature by removal of the dry ice-acetone bath. In a second flask, a 100-mL, two-necked, round-bottomed flask, equipped with a magnetic stirring bar, rubber septum, and a gas-flow adapter with a stopcock, are placed 74.7 mg (0.120 mmol) of (R)-2,2'-bis(diphenylphosphino)-1,1'-binaphthyl [(R)-BINAP] (Notes 3, 4), and 25.8 mg (0.100 mmol) of acetylacetonatobis(ethylene)rhodium(I) (Note 5). After the flask is flushed with nitrogen, 50 mL of 1,4-dioxane is added via syringe, and the mixture is stirred at room temperature for 10 min. To the first flask containing the phenylborate reagent, are added successively, via syringe 0.975 g (10.1 mmol) of 2-cyclohexenone (Note 6), 0.54 mL of water, and the catalyst solution from the second flask. The entire mixture is heated in an oil bath at 100°C for 12 hr. After the solvent is cooled to room temperature, it is removed under reduced pressure on a rotary evaporator. The dark brown residue is diluted with diethyl ether (100 mL) and transferred to a 500-mL separatory funnel. The ether solution is washed with 10% hydrochloric acid (100 mL) and 5% aqueous sodium hydroxide (100 mL). The aqueous layers are extracted with diethyl ether (30 mL each). The organic layers are combined, washed with aqueous saturated sodium chloride (100 mL), and dried over anhydrous magnesium sulfate. The solvent is removed under reduced pressure on a rotary evaporator to give a brown oil, which is dissolved in 5 mL of diethyl ether and chromatographed on silica gel [4 cm Φ, 200 mL of silica gel (Note 14)]. Elution with 500 mL of hexane (Note 15) followed by elution with 400 mL of diethyl ether gives crude 3-phenylcyclohexanone (1), collected as 20-mL fractions and combined. Distillation under reduced pressure (bp 125–130°C at 0.5 mm) gives 1.55 g (88.1% yield) of (R)-3-phenylcyclohexanone (1) (Note 8) as a colorless oil. The enantiomeric purity is 98.3% ee, which is determined by HPLC analysis using a chiral stationary phase column (Note 9).

2. Notes

1. Phenylboronic acid was purchased from Tokyo Kasei Kogyo Co., Ltd. and used as received.

2. The use of 2.5 equiv (to 2-cyclohexenone) of phenylboronic acid is important for a high yield. With 1.0 equiv of phenylboronic acid, the yield is less than 70%.

3. (R)-BINAP is commercially available from Aldrich Chemical Company, Inc., although the submitters have prepared it according to the reported procedures.[2]

4. This molar ratio (1.2 : 1) of BINAP to the rhodium(I) is required for high enantioselectivity. With a 1 : 1 ratio, the enantioselectivity is usually 1 or 2% lower.

5. Acetylacetonatobis(ethylene)rhodium(I) is commercially available from Strem Chemicals, Inc., although the submitters have prepared it according to a reported procedure.[3]

6. 2-Cyclohexenone was purchased from Tokyo Kasei Kogyo Co., Ltd. and distilled before use.

7. 1,4-Dioxane was purchased from Wako Pure Chemical Industries, Ltd. and distilled from benzophenone ketyl before use.

8. Specific rotation value of **1**: $[\alpha]_D^{20}$ +21° (CHCl$_3$, c 0.96) [literature rotation for (R)-**1** (98.7% ee);[4] $[\alpha]_D^{20}$ +20.5° (c 0.58, CHCl$_3$)]. The spectra are as follows: ^{1}H NMR (500 MHz, CDCl$_3$) δ: 1.77 (qdd, 1 H, J = 12.6, 4.4, 3.3), 1.85 (qd, 1 H, J = 12.3, 3.3), 2.07 (dm, 1 H, J = 13.1), 2.14 (ddq, 1 H, J = 12.7, 6.0, 3.1), 2.37 (tdd, 1 H, J = 12.6, 6.3, 1.1), 2.45 (dm, 1 H, J = 14.5), 2.52 (td, 1 H, J = 12.4, 1.1), 2.59 (ddt, 1 H, J = 14.0, 4.5, 2.0), 3.00 (tt, 1 H, J = 11.9, 3.8), 7.19-7.25 (m, 3 H), 7.32 (t, 2 H, J = 7.5); ^{13}C NMR (125 MHz, CDCl$_3$) δ: 24.88, 32.10, 40.46, 44.04, 48.21, 126.01, 126.03, 128.06, 143.86, 209.80.

9. The column contained Daicel Chiralcel OD-H (eluent, hexane/2-propanol = 98/2).

10. Bromobenzene was purchased from Kanto Chemical Company Inc., and used as received.

11. Diethyl ether was purchased from Nacalai Tesque Company Inc., and distilled from benzophenone ketyl before use.

12. n-Butyllithium in hexane was purchased from Kanto Chemical Company Inc., and used as received.

13. Trimethoxyborane (trimethyl borate) was purchased from Wako Pure Chemical Industries, Ltd. and used as received.

14. Silica gel 60 100–210 μm (Kanto Chemical Company Inc.) was used.

15. Biphenyl as a by-product is removed by this elution.

3. Discussion

Conjugate addition of organometallic reagents to electron-deficient olefins constitutes one of the versatile methodologies for forming carbon-carbon bonds. Although considerable efforts have been made to develop efficient chiral catalytic systems for asymmetric conjugate addition, the successful examples are rare in terms of enantioselectivity, catalytic activity, and generality.[5] In 1997, Miyaura and co-workers[6] found that a phosphine-rhodium complex catalyzes the 1,4-addition of aryl- and alkenylboronic acids to α,β-unsaturated ketones giving β-substituted ketones. Based on this finding, the submitters developed rhodium-catalyzed, asymmetric, 1,4-addition reactions of organoboron reagents to electron-deficient olefins. High enantioselectivity has been achieved in the reaction of α,β-unsaturated ketones with aryl- and alkenylboronic acids, which was carried out in dioxane/H$_2$O at 100°C in the presence of a rhodium catalyst generated from acetylacetonatobis(ethylene)rhodium(I) and (S)-2,2'-bis(diphenylphosphino)-1,1'-binaphthyl (BINAP).[7] Both acyclic and cyclic enones gave the corresponding optically active ketones of over 90% ee. In place of isolated organoboronic acids, 2-alkenyl-1,3,2-benzodioxaboroles, readily accessible by hydroboration of alkynes with catecholborane,[8] and arylborates, generated by reaction of aryllithiums with trimethoxyborane,[9] can also be used for the asymmetric 1,4-addition. α,β-Unsaturated esters[10] and 1-alkenylphosphonates[11] undergo the rhodium-catalyzed asymmetric addition of organoboron derivatives to give the corresponding 1,4-addition products of over 90% ee. The submitters also studied some basic factors, including reaction temperature, solvent, rhodium precursor, and chiral ligand, which are capable of affecting the enantioselectivity and catalytic activity.[12] Some of the optically active products obtained by the rhodium-catalyzed, asymmetric, 1,4-addition of aryl- and alkenylboronic acids or their derivatives are shown in Figure 1.

Figure 1. Products obtained by the rhodium-catalyzed asymmetric 1,4-additiion

[(S)-BINAP is used]

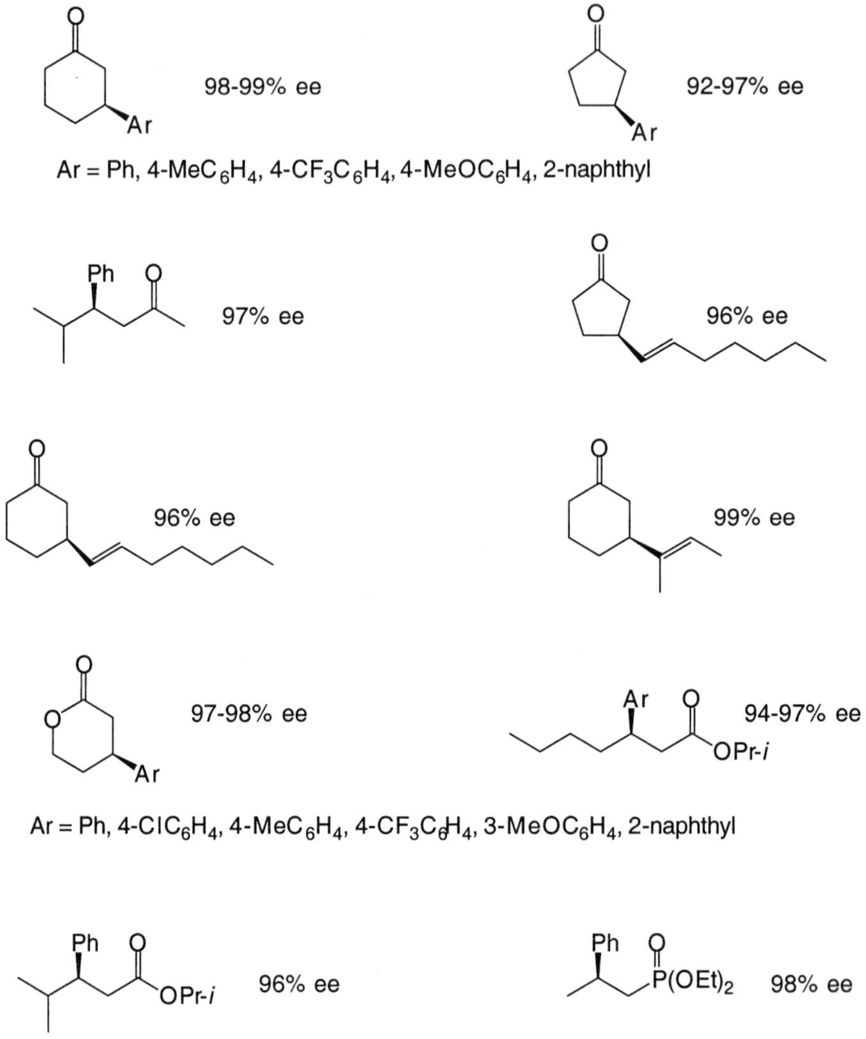

98-99% ee

92-97% ee

Ar = Ph, 4-MeC$_6$H$_4$, 4-CF$_3$C$_6$H$_4$, 4-MeOC$_6$H$_4$, 2-naphthyl

97% ee

96% ee

96% ee

99% ee

97-98% ee

94-97% ee

Ar = Ph, 4-ClC$_6$H$_4$, 4-MeC$_6$H$_4$, 4-CF$_3$C$_6$H$_4$, 3-MeOC$_6$H$_4$, 2-naphthyl

96% ee

98% ee

1. Department of Chemistry, Faculty of Science, Kyoto University, Sakyo, Kyoto 606-8502, Japan

2. Cai, D.; Payack, J. F.; Bender, D. R.; Hughes, D. L.; Verhoeven, T. R.; Reider, P. J. *J. Org. Chem.* **1994**, *59*, 7180.

3. Cramer, R. *Inorg. Synth.* **1974**, *15*, 14.

4. Schultz, A. G.; Harrington, R. E. *J. Am. Chem. Soc.* **1991**, *113*, 4926.

5. For a pertinent review, see Tomioka, K.; Nagaoka, Y. In *Comprehensive Asymmetric Catalysis;* Jacobsen, E. N.; Pfaltz, A.; Yamamoto, H. (Eds.); Springer, **1999**; *3,* pp 1105-1120.

6. Sakai, M.; Hayashi, H.; Miyaura, N. *Organometallics* **1997**, *16*, 4229.

7. For leading references, see Takaya, Y.; Ogasawara, M.; Hayashi, T.; Sakai, M.; Miyaura, N. *J. Am. Chem. Soc.* **1998**, *120*, 5579.

8. Takaya, Y.; Ogasawara, M.; Hayashi, T. *Tetrahedron Lett.* **1998**, *39*, 8479.

9. Takaya, Y.; Ogasawara, M.; Hayashi, T. *Tetrahedron Lett.* **1999**, *40*, 6957.

10. Takaya, Y.; Senda, T.; Kurushima, H.; Ogasawara, M.; Hayashi, T. *Tetrahedron: Asymmetry* **1999**, *10*, 4047.

11. Hayashi, T.; Senda, T.; Takaya, Y.; Ogasawara, M. *J. Am. Chem. Soc.* **1999**, *121*, 11591.

12. Takaya, Y.; Ogasawara, M.; Hayashi, T. Chirality **2000**, *12*, 469.

Appendix

Chemical Abstracts Nomenclature (Collective Index Number); (Registry Number)

(R)-3-Phenylcyclohexanone: Cyclohexanone, 3-phenyl-, (R)- (9); (34993-51-6)

Phenylboronic acid: Benzeneboronic acid (8); Boronic acid, phenyl- (9); (98-80-6)

(R)-2,2'-Bis(diphenylphosphino)-1,1'-binaphthyl [(R)-BINAP]: Phosphine oxide, [1,1'-binaphthalene]-2,2'-diylbis[diphenyl-, (R)- (11); (94041-16-4)

Acetylacetonatobis(ethylene)rhodium (1): Rhodium, bis(ethylene)(2,4-pentanedionato)- (8); Rhodium, bis(η^2-ethene)(2,4-pentanedionato-O,O'- (9); (12082-47-2)

Cyclohexenone: HIGHLY TOXIC: 2-Cyclohexen-1-one (8,9); (930-68-7)

Bromobenzene: Benzene, bromo- (8,9); (108-86-1)

Butyllithium: Lithium, butyl-(8,9); (109-72-8)

Trimethoxyborane: ALDRICH; Trimethyl borate: Boric acid, trimethyl ester (8,9); (121-43-7)

SYNTHESIS OF (+)-(1S,2R)- AND (-)-(1R,2S)-trans-2-PHENYLCYCLOHEXANOL VIA SHARPLESS ASYMMETRIC DIHYDROXYLATION (AD)

[Cyclohexanol, 2-phenyl-, (1S-trans)- and Cyclohexanol, 2-phenyl-, (1R-trans)-]

Submitted by Javier Gonzalez, Christine Aurigemma, and Larry Truesdale.[1]

Checked by Scott E. Denmark, Steven A. Tymonko, Jeromy J. Cottell, and Laurent Gomez.

1. Procedure

A. *(+)-(1R,2R)-1-Phenylcyclohexane-cis-1,2-diol* (**2**). A 3-L flask with a mechanical stirrer, thermometer and an inlet port open to the atmosphere is charged with 375 mL of water. Stirring is started and the following reagents are added in the order indicated through a powder funnel: potassium ferricyanide (247 g, 0.75 mol, 3 equiv), anhydrous potassium carbonate (104 g, 0.75 mol, 3 equiv), methanesulfonamide (23.8 g, 0.25 mol, 1 equiv), potassium osmate dihydrate (46.1 mg, 0.125 mmol, 0.05 mol %), (DHQD)$_2$PHAL [1,4-bis(9-O-dihydroquinidinyl)phthalazine, 486.9 mg, 0.625 mmol, 0.25 mol %], 1-phenylcyclohexene (**1**, 39.55 g, 0.25 mol) and tert-butyl alcohol (250 mL) (Notes 1 and 2).

The slurry is stirred vigorously for 2 days at a rate of 500 rpm. During this time, the product crystallizes in the top organic phase, beginning after 4 hr. Also, the appearance of the slurry gradually changes from a mixture containing red granules (ferricyanide) to yellow flakes, which are presumably a salt of iron(II).

After the reaction is complete the mixture is treated with ethyl acetate (250 mL) with stirring to dissolve the product. The resulting mixture is filtered through a 500-mL medium-fritted glass funnel and the flask and filter cake are washed with ethyl acetate (3 x 50 mL). The filtrate is transferred to a 2-L separatory funnel and the brown-colored aqueous phase is separated. The organic phase is washed with 2 M potassium hydroxide (KOH, 2 x 50 mL) with vigorous shaking to remove methanesulfonamide, then dried over magnesium sulfate (MgSO$_4$, 12.5 g). The solid is filtered, the cake is washed with ethyl acetate (2 x 37 mL) and the filtrate is evaporated, to afford a white solid. After the crude product is dried under reduced pressure overnight, it weighs 47.44 g (99%) (Notes 3 and 4).

B. *(-)-(1R,2S)-trans-2-Phenyl-1-cyclohexanol* (**3**). A 1-L, three-necked flask is set up with a mechanical stirrer, thermometer, reflux condenser and nitrogen line. The flask is placed in a silicone oil bath (230 x 100-mm Schott crystallizing dish). The flask is charged with a slurry containing activated W-2 Raney nickel (257.5 g) in wet ethanol (70% v/v) through a powder funnel under a blanket of nitrogen. *Caution-Fire Hazard! Raney nickel is extremely pyrophoric when dry and must be kept submerged under liquid at all times* (Note 5). This is done by transferring an aqueous slurry of Raney nickel to the flask with the aid of 250 mL of anhydrous ethanol in portions, making sure that the catalyst does not dry. The crude diol from the previous step is added to the flask, using another 50 mL of anhydrous ethanol to complete the transfer. The mixture is stirred vigorously and refluxed for 2 hr (Note 6).

The reaction mixture is allowed to cool to 40-50°C and filtered through a 1/2"-layer of Celite in a 500-mL fritted (medium) glass funnel, making sure that the liquid level does not fall below the surface of the filter cake. A total of 300 mL of ethanol in small portions is used to transfer the slurry quantitatively to the funnel and to wash the filter cake. A 170-mm x 90-mm crystallizing dish is useful to cover the top of funnel during the filtration; however it is not necessary. The Raney nickel sludge is transferred with water to a waste container.

The filtrate is concentrated under reduced pressure to remove most of the ethanol and the resulting two-phase mixture is diluted with 50 mL of saturated brine and extracted with ethyl acetate (2 x 50 mL). The organic phase is dried over $MgSO_4$ (3.75 g), filtered and evaporated under reduced pressure overnight to give the crude alcohol (35.53 g, 94.0% ee). The solid residue is crystallized from 75 mL of warm petroleum ether (bp 30-60°C). The crystallization is allowed to proceed for 3 hr at room temperature and 1 hr at 0°C. The crystalline mass is triturated with 25 mL of cold (0°C) pentane to break up the lumps. The slurry is filtered, washed with chilled pentane in portions (3 x 12.5 mL) and the solid is dried under reduced pressure overnight to constant weight, to afford 25.12 g (58% from 1) of colorless crystals, mp 64-65°C. The enantiomeric purity was found to be >99.5% ee by chiral stationary phase SFC (supercritical-fluid chromatography) (Notes 7-8).

2. Notes

1. The potassium ferricyanide was Spectrum technical grade. Methane-sulfonamide (97%), $(DHQ)_2PHAL$ (95%), potassium osmate dihydrate and tert-butyl alcohol (99.3%) were obtained from Aldrich Chemical Company, Inc. 1-Phenylcyclohexene (98%) was obtained from Acros Organics.

2. The temperature did not rise more than a few degrees above room temperature after all the reagents were added.

3. This material, which is contaminated with the chiral ligand, is sufficiently pure to be used in the next step. A small portion of the crude diol (479 mg) was purified by column chromatography (SiO_2, 30 x 160 mm, 4/1 hexane/ ethyl acetate) to yield a white solid (430 mg, 90%). The enantiomeric purity of the crude product was determined to be 99.4% ee

by SFC analysis on a chiral stationary phase using a Berger Supercritical Fluid Chromatograph (Berger Instruments, Newark, DE). The diols were separated using an (R,R) Whelk-O, 250 mm x 4.6 mm, 5μ, 100 Å column at 40°C. Methanol was used as the modifier and held isocratically at 15% in CO_2. The eluent flow rate was maintained at 2.0 mL/min and the outlet pressure was isobaric at 150 bar (112,500mm, 11.3×10^4mm). All samples were prepared in methanol at concentrations of 1 mg/mL. The retention time for the diol peaks were 3.55 min for the (+)-isomer (**2**) and 3.05 min for the (-)-isomer. The enantiomeric purity of the crude product was also analyzed by HPLC using an (R.R) Whelk-O, 250 mm x 4.6 mm, 5μ, 100 Å column at 150 bar (112, 500mm). Isopropyl alcohol was used as the modifier and held isocratically at 15% in hexane. The eluent flow rate was maintained at 1.3 mL/min, and the retention time for the diol peaks were 7.56 min for the (+)-isomer (**2**) and 5.30 min for the (-)-isomer.

The (-)-enantiomer was prepared using this procedure with $(DHQ)_2PHAL$ as the chiral ligand. The yield of the crude diol, (-)-(1S,2S)-1-phenylcyclohexane-cis-1,2-diol, was 47.39 g (99%). The enantiomeric purity of this material was found to be 98.9% ee by CSP-SFC (see Note 4 for data). These levels of asymmetric induction are similar to those reported by Sharpless and co-workers, i.e., 99% ee for the (R,R)-isomer using AD-mix-β [containing $(DHQD)_2PHAL$] and 97% ee for the (S,S)-isomer using AD-mix-α [containing $(DHQ)_2PHAL$].[2]

4. The diol was further purified by recrystallization from ethyl acetate and petroleum ether in 62% recovery. The (+)-(1R,2R)-1-phenylcyclohexane-cis-1,2-diol thus obtained had a mp of 122-123°C (lit. mp 121-122°C)[3] and was found to have an optical purity of >99.5% ee by chiral SFC. The following physical data was observed:

(+)-(1R,2R)-1-Phenylcyclohexane-1,2-diol. ^{1}H NMR (500 M, $CDCl_3$) δ: 1.36-1.45 (m, 1 H), 1.52- 1.56 (m, 1 H), 1.65- 1.75 (m, 4 H), 1.83- 1.90 (m, 3 H), 2.69 (d, 1 H, J = 2.0), 3.95 (dt, 1 H, J = 4.0, 11.0), 7.27 (t, 1 H, J = 7.5), 7.38 (t, 2 H, J = 7.5), 7.50 (d, 2 H, J = 8.5); ^{13}C NMR (125 M, $CDCl_3$) δ: 21.27, 24.58, 29.40, 38.73, 74.74, 76.01,

125.40, 127.24, 128.67, 146.67; IR (KBr)cm^{-1}: 3480, 3241, 3083, 3056, 3021, 2948, 2859, 1945, 1866, 1797, 1741, 1598, 1494, 1444, 1411, 1375, 1268, 1203, 1141, 1079, 1064, 991, 968, 754, 694; LRMS EI (relative intensity) 192 (53), 174 (13), 145 (8), 133 (100), 120 (40), 105 (56), 91 (16), 77 (31), 70 (14), 55 (23); $[\alpha]_D$ +16.8° (benzene, c 1.0) Calcd. for $C_{12}H_{16}O_2$: C; 74.97, H; 8.39. Found: C; 74.90, H; 8.62.

(-)-(1S,2S)-1-Phenylcyclohexane-1,2-diol. mp 122-123°C; $[\alpha]_D$ −16.0° (benzene, c 1.0,); Calcd. for $C_{12}H_{16}O_2$: C; 74.97, H; 8.39. Found: C; 74.89, H; 8.62.

5. The Raney nickel (WR Grace Grade 28) was obtained as a 50 wt% aqueous slurry from Strem Chemicals Inc. The mass of the Raney nickel was determined by the following procedure:[4] The weight, in grams, of a 500-mL volumetric flask filled with deionized water was recorded (Mass A). A portion of the water was removed and replaced with the Raney Nickel slurry. The remaining volume was filled with deionized water and reweighed (Mass B). The amount of Raney nickel, in grams, was calculated using the equation Amt. = 1.167(Mass B - Mass A), where 1.167 accounts for the volume of water displaced by the Raney nickel catalyst with an average density of 7.00 g/mL. However, prior to transferring to the flask, the excess water was decanted from the material. Small spills of Raney nickel slurry were transferred with a wet Kimwipe to a waste container containing water. The ethanol was undenatured.

6. The progress of the reaction may be followed by TLC using 1:3 ethyl acetate/petroleum ether and visualization with anisaldehyde stain. The diol appears at an R_f = ~0.3 (stains olive-brown); the 2-phenylcyclohexanol appears at an R_f = ~0.6 (stains blue). The checkers found that the use of a needle outlet at the top of the condenser helped to maintain a smooth reflux and prevented leakage though the stirring assembly.

7. The enantiomeric purity was determined by CSP-SFC (Note 3). Samples were run isocratically at 1% methanol-modified CO_2 and the flow rate was held constant at 4.2 mL/min. Outlet pressure was isobaric at 150 bar (112,500mm, 11.3 x 10^{4}mm). All samples were prepared in methanol at concentrations of 1 mg/mL. The retention times for

97

the enantiomers were 4.50 min for the (-)-isomer and 3.91 min for the (+)-isomer. The enantiomeric purity of the alcohol was also analyzed by HPLC using an (R.R) Whelk-O, 250 mm x 4.6 mm, 5μ, 100 Å column. Isopropyl alcohol was used as the modifier and held isocratically at 2% in hexane. The eluent flow rate was maintained at 1.3 mL/min, and the retention times for the diol peaks were 14.05 min for the (-)-isomer (3) and 11.46 min for the (+)-isomer.

Sublimation of a small portion of the alcohol (501 mg) at 45°C (0.05 mm) provided a white powder (467 mg, 93%), which was found to be analytically pure. The following physical data were observed:

(-)-(1R,2S)-2-phenyl-1-cyclohexanol. ^{1}H NMR (500 MHz, CDCl$_3$) δ: 1.31-1.57 (m, 4 H), 1.64 (d, 1 H, J = 2.5), 1.78-1.80 (m, 1 H), 1.86-1.89 (m, 2 H), 2.12- 2.14 (m, 1 H), 2.42-2.47 (m, 1 H), 3.64-3.69 (m, 1 H), 7.24-7.28 (m, 3 H), 7.35 (t, 3 H, J = 7.5); ^{13}C NMR (125 MHz, CDCl$_3$) δ: 25.28, 26.27, 33.53, 34.64, 53.43, 74.63, 127.07, 128.18, 129.01, 143.57; IR (KBr) cm[-1]: 3295, 3081, 3058, 3025, 2854, 2653, 1945, 1878, 1801, 1756, 1600, 1492, 1446, 1348, 1060, 964, 1130, 746, 696; LRMS EI (relative intensity) 176.1 (61), 158.1 (9), 143.1 (11), 130.1 (47), 117.1 (39), 104.1 (43), 91.0 (100), 85.1 (7), 77.0 (16), 71.0 (7), 65.0 (10), 57.0 (17); [α]$_D$ −58.3° (MeOH ,c 1.0); Calcd. for C$_{12}$H$_{16}$O: C; 81.77, H; 9.15. Found: C; 81.96, H; 9.28.

(+)-(1S,2R)-2-phenyl-1-cyclohexanol. mp 64-66°C; [α]$_D$ +59.4° (MeOH, c 1.0); Calcd. for C$_{12}$H$_{16}$O: C; 81.77, H; 9.15. Found: C; 81.89, H; 9.23

A second batch of product was obtained by concentration and recrystallization of the mother liquor to provide another 1.89 g (4.4%) of (-)-(1R,2S)-trans-2-phenyl-1-cyclohexanol of a slightly lower enantiomeric purity (98.6% ee).

8. The other enantiomer, (+)-(1S,2R)-trans-2-phenylcyclohexanol, was also prepared by this procedure. The crude mass of the alcohol was 39.15 g (95.8% ee), while after recrystallization, the yield was 29.80 g (69%) of material with the following properties: mp 64-66°C, >99.5% ee (see Note 7 for data). The reaction appears to be scaleable

since the submitters reported obtaining a 69% overall yield (122 g) of the (-) isomer from 1 when this procedure was carried out on a 1-mol scale.

Waste Disposal Information

All toxic materials were disposed of in accordance with "Prudent Practices in the Laboratory"; National Academy Press; Washington, DC, 1995.

3. Discussion

Enantiomerically pure trans-2-phenylcyclohexanol, first used by Whitesell as a chiral auxiliary[5] has become a popular reagent in a number of asymmetric transformations.[6] Some recent applications include asymmetric azo-ene reactions,[7] [4 + 2]-cycloaddition reactions,[8] ketene-olefin [2 + 2]-reactions,[9] enolate-imine cyclocondensations,[10] Pauson-Khand reactions,[11] palladium annulations[12] and Reformatsky reactions.[13] Despite its potential, use of this chiral auxiliary on a preparative scale is currently limited by its prohibitive cost.

A previous *Organic Syntheses* procedure employing Whitesell's method affords both enantiomers of trans-2-phenylcyclohexanol in 3-4 steps from cyclohexene oxide via a lipase-catalyzed hydrolysis of the corresponding racemic chloroacetate ester.[14] A related reaction, the Lipase PS30-catalyzed kinetic acetylation of the racemic alcohol afforded the enantiomers in excellent yield and optical purity.[15] However, a limitation of both of these procedures is that to obtain one of the enantiomers, they require chromatographic purification of an intermediate, making scale-up impractical. The (+)-enantiomer (≥97% ee) has been prepared from 1-phenyl-1-cyclohexene by hydroboration with $IpcBH_2$, monoisopinocamphenylborane, but the reaction is slow, resulting in 70% conversion after 7 days at 0°C.[16] A procedure involving ring opening of cyclohexene oxide with phenyllithium in the presence of a chiral additive has been reported, but the level of asymmetric induction is modest (47% ee).[17] Other methods that have been used recently to prepare this chiral auxiliary involve enantioselective epoxidation[18] and protonation.[19]

The preparation described here is a slight modification of a route published by King and Sharpless[3] via the osmium-catalyzed asymmetric dihydroxylation (AD) reaction of 1-phenyl-1-cyclohexene.[2] The major strengths of this process are that either enantiomer can be prepared in high optical purity (> 99.5% ee) without the need for chromatography.

Some experimentation afforded improvements to the process. For example, in the case of the AD reaction, both the osmium and chiral concentrations could be reduced to a level of 0.05 mol % and 0.25 mol %, respectively, or one-fourth the levels in the commercial AD-mix formulation, without compromising the yield and enantiomeric excess of the crude product. The volume of liquid was also reduced to one-fourth of the quantities reported (1.5 L of water and 1 L of tert-butyl alcohol per mole of substrate versus 5 L of water and 5 L of tert-butyl alcohol per mole of substrate). Under these conditions the reaction mixture is a slurry, but the potassium ferricyanide dissolves as it reacts. Reducing the catalyst concentration had the effect of doubling the reaction time from 1 day to 2 days. Interestingly, a study on the use of reduced amounts of osmium in the AD reaction of 1-phenyl-1-cyclohexene concluded that reducing the quantity of osmium by half (to 0.1 mol %), doubled the reaction time without affecting the yield, but that further reductions in osmium content made the reaction too sluggish to be useful.[20]

In scaling up this procedure, the biggest improvement in the overall yield was achieved by omitting the crystallization of the intermediate diol. The trans-2-phenylcyclohexanol, which forms relatively large crystals, is easier to handle than the diol, which is a very fluffy powder. Analysis of the final product was carried out by both CSP-HPLC and CSP-SFC methods.

1.	Combinatorial Chemistry Technology Department, High-Throughput Discovery, Pfizer, Global Research and Development, La Jolla, CA.

2.	Sharpless, K. B.; Amberg, W.; Bennani, Y. L.; Crispino, G. A.; Hartung, J.; Jeong, K. -S.; Kwong, H. -L.; Morikawa, K.; Wang, Z. -M.; Xu, D.; Zhang, X. -L. *J. Org. Chem.* **1992**, *57*, 2768.

3.	King, S. B.; Sharpless, K. B. *Tetrahedron Lett.* **1994**, *35*, 5611.

4.	Activated Metals & Chemicals, Inc., P.O. Box 4130, Sevierville, TN 37864

5.	Whitesell, J. K.; Chen, H. -H.; Lawrence, R. M. *J. Org. Chem.* **1985**, *50*, 4663.

6.	For an early review, see: Whitesell, J. K. *Chem. Rev.* **1992**, *92*, 953.

7.	Brimble, M. A.; Lee, C. K. Y. *Tetrahedron:Asymmetry* **1998**, *9*, 873.

8.	(a) Denmark, S. E.; Hurd, A. R.; Sach, H. J. *J. Org. Chem.* **1997**, *62*, 1668; (b) Brimble, M. A.; Mcewan, J. F.; Turner, P. *Tetrahedron: Asymmetry* **1998**, *9*, 1239.

9.	De Azevedo, M. B. M.; Murta, M. M.; Greene, A. E. *J. Org. Chem.* **1992**, *57*, 4567.

10.	Ojima, I.; Habus, I.; Zhao, M.; Zucco, M.; Park, Y. H.; Sun, C. M.; Brigaud, T. *Tetrahedron* **1992**, *48*, 6985.

11.	Castro, J.; Moyano, A.; Pericàs, M. A.; Riera, A.; Greene, A. E.; Alvarez-Larena, A.; Piniella, J. F. *J. Org. Chem.* **1996**, *61*, 9016.

12.	Chassaing, C.; Haudrechy, A.; Langlois, Y. *Tetrahedron Lett.* **1999**, *40*, 8805.

13.	Shankar, B. B.; Kirkup, M. P.; McCombie, S. W.; Clader, J. W.; Ganguly, A. K. *Tetrahedron Lett.* **1996**, *37*, 4095.

14.	Schwartz, A.; Madan, P.; Whitesell, J. K.; Lawrence, R. M. *Org. Synth., Coll. Vol. VIII* **1993**, 516.

15.	Carpenter, B. E.; Hunt, I. R.; Keay, B. A. *Tetrahedron:Asymmetry* **1996**, *7*, 3107.

16.	Brown, H. C.; Vara Prasada, J. V. N.; Gupta, A. K.; Bakshi, R. K. *J. Org. Chem.* **1987**, *52*, 310.

17.	Mizuno, M.; Kanai, M.; Iida, A.; Tomioka, K. *Tetrahedron* **1997**, *53*, 10699.

18. Brandes, B. D.; Jacobsen, E. N. *J. Org. Chem.* **1994**, *59*, 4378.

19. Asensio, G.; Gaviña, A. C. P.; Medio-Simon, M. *Tetrahedron Lett.* **1999**, *40*, 3939.

20. McIntosh, J. M.; Kiser, E. J. *Synlett* **1997**, 1283.

Appendix

Chemical Abstracts Nomenclature (Collective Index Number);

(Registry Number)

(+)-(1S,2R)-trans-2-Phenylcyclohexanol: Cyclohexanol, 2-phenyl-, (1S-trans)- (9); (34281-92-0)

(-)-(1R,2S)-trans-2-Phenylcyclohexanol: Cyclohexanol, 2-phenyl-, (1R-trans)- (11); (98919-68-7)

(+)-(1R,2R)-1-Phenylcyclohexane-cis-1,2-diol: 1,2-Cyclohexanediol, 1-phenyl-, (1R-cis)- (12); (125132-75-4)

Potassium ferricyanide: Ferrate (3-), hexacyano-, tripotassium (8); Ferrate (3-), hexakis(cyano-C)-, tripotassium, (OC-6-11)- (9); (13746-66-2)

Methanesulfonamide (8, 9); (3144-09-0)

Potassium osmate (VI) dihydrate: Osmic acid, dipotassium salt (8, 9); (19718-36-3)

(DHQ)$_2$PHAL: ALDRICH: Hydroquinine, 1,4-phthalazinediyl diether: Cinchonan, 9,9"-[1,4-phthalazinediylbis(oxy)bis[10,11-dihydro-6'-methoxy-, (8α,9R)-(8"α,9"R)- (13); (140924-50-1)

1-Phenylcyclohexene: Benzene, 1-cyclohexen-1-yl- (8, 9); (771-98-2)

tert-Butyl alcohol (8); 2-Propanol, 2-methyl- (9); (75-65-0)

Raney nickel: CANCER SUSPECT AGENT: Nickel (8, 9); (7440-02-0)

(-)-(1S,2S)-1-Phenylcyclohexane-cis-1,2-diol: 1,2-Cyclohexanediol, 1-phenyl-, (1S-cis)- (9); (34281-90-8)

DICYCLOHEXYLBORON TRIFLUOROMETHANESULFONATE

[Methanesulfonic acid, trifluoro-, anhydride with dicyclohexylborinic acid]

Submitted by Atsushi Abiko.[1]

Checked by Eric M. Flamme and William R. Roush.

1. Procedure

Caution! Borane-dimethyl sulfide complex is foul smelling. All operations using dimethyl sulfide must be carried out in a well-ventilated hood.

Dicyclohexylboron trifluoromethanesulfonate.[2] An oven-dried, 250-mL, round-bottomed flask containing a stir bar and capped with a rubber septum is charged with cyclohexene (33.4 mL, 0.33 mol) (Note 1) and dry diethyl ether (100 mL) (Note 2), and kept at 0°C under nitrogen. Borane-dimethyl sulfide complex (16.6 mL, 0.16 mol) (Note 1) is added dropwise during 30 min with stirring. The reaction mixture is stirred for 3 hr at 0°C, then the solid is allowed to settle without stirring. The supernatant organic solution is removed as much as possible by syringe (Notes 3, 4), and the residual solid is dried under reduced pressure (Note 5) to give dicyclohexylborane (26.3-28.3 g, 92-99% yield) (Note 6), which is used without purification for the preparation of the triflate.

The solid is suspended in 100 mL of dry hexane (Note 2) and trifluoromethanesulfonic acid (24.0 g, 0.16 mol) (Note 1) is added dropwise via a glass syringe (Notes 7, 8) over 30 min with constant stirring, during which time vigorous gas evolution occurs; the solid gradually dissolves, and the solution develops a yellow-orange color (Note 9). Stirring is continued at room temperature for 1 hr, then the reaction mixture is left for 1-2 hr without stirring. A semi-solid phase separates and the top layer is transferred via cannula into a dry, 250 mL round-bottomed flask (Note 10). The flask is placed in a -20°C freezer for 36 hr. Large crystals form and the mother liquor is transferred via cannula to another dry, 100-mL, round-bottomed flask. The crystals are dried under reduced pressure at 0°C for 5 hr giving 40.8 g (78%) of dicyclohexylboron trifluoromethanesulfonate. The mother liquors are concentrated to dryness under nitrogen (N_2) using a rotary evaporator (Note 11). The residue is redissolved in 20 mL of dry hexane and crystallized in a -20°C freezer as described above to give an additional 7 g of product. The total yield of dicyclohexylboron trifluoromethanesulfonate is 47.8 g (92%) (Notes 12, 13). The isolated crystals are dissolved in hexane in a graduated cylinder equipped with a ground glass joint and a stopcock-equipped syringe inlet to make a 1 M stock solution (Notes 14, 15).

2. Notes

1. Cyclohexene was purchased from Wako Pure Chemical Ltd. Japan, or Aldrich Chemical Company, Inc., and used after distillation from lithium aluminum hydride. Borane-dimethyl sulfide complex was obtained from Aldrich Chemical Company, Inc., and was used as received.[3] Trifluoromethanesulfonic acid was purchased from Wako Pure Chemical Ltd. Japan or Aldrich Chemical Company, Inc., and used without purification. The checkers used a freshly opened ampule of trifluoromethanesulfonic acid for each run.

2. Diethyl ether was freshly distilled from benzophenone ketyl. Hexane was distilled from lithium aluminum hydride.

3. The supernatant organic solution might contain active borane. The solution that was removed was treated with methanol to destroy any active borane. The waste containing dimethyl sulfide must be treated appropriately before being discarded.

4. Care must be taken to avoid loss of product that is easily taken up in the syringe. The checkers found it more convenient to perform the reaction in a tared, round-bottomed Schlenk flask. Solvent was then removed from the solid dicyclohexylborane by filtration using a positive nitrogen flow.

5. Dicyclohexylborane bumps during the drying step, and the checkers found that it is advantageous to use a vacuum adapter containing a glass frit and an auxiliary cold trap to prevent contamination of the vacuum line with the product.

6. The submitter obtained 27.2-30.0 g (96-105% yield) of dicyclohexylborane.

7. Trifluoromethanesulfonic acid is highly corrosive, and a glass syringe must be used for this operation. Plastic syringes are rapidly destroyed by trifluoromethanesulfonic acid, leading to safety hazards if used for this procedure.

8. The reaction of dicyclohexylborane and trifluoromethanesulfonic acid is highly exothermic. On one occasion, the checkers cooled the reaction in an ice bath during the addition period, with no effect on product yield. The submitter reports that he once experienced a sudden vigorous reaction under cooling conditions, probably due to accumulation of unreacted triflic acid. It thus appears safer to add the acid at room temperature, slowly, so that it reacts immediately.

9. The submitter obtained a colorless solution of boron triflate, but the checkers observed development of a yellow-orange color upon addition of triflic acid. Regardless of the color, the triflate solution could be used without a decrease in yields.

10. The small semi-solid yellow layer (about 2 mL) was left behind.

11. The rotary evaporator was back-flushed with dry N_2, and it is recommended that a drying column be inserted between the evaporator and the water aspirator.

12. Dicyclohexylboron trifluoromethanesulfonate exhibited the following spectroscopic properties: ^{1}H NMR (400 MHz, CDCl$_3$) δ: 67 (m, 10 H), 1.19 (m, 12 H); ^{13}C NMR (100 MHz, CDCl$_3$) δ: 118.3 (q, J = 316), 30.5, 26.9, 26.6, 26.5; ^{11}B NMR (115 MHz, hexane/C$_6$D$_6$: δ relative to BF$_3$:Et$_2$O) δ 59.2 (br s); lit.[2b] ^{11}B NMR δ 59.6 (br s).

13. The submitter obtained the product in 89% yield (one crop) by using this crystallization procedure.

14. An alternative isolation procedure follows. When the reaction between dicyclohexylborane and trifluoromethanesulfonic acid was complete, the liquid layer was transferred via cannula to a tared, graduated cylinder, the top of which had been modified with a stopcock-equipped syringe inlet and a ground glass joint. This flask was directly attached to a rotary evaporator (Note 10) for removal of solvent. The resulting orange-yellow oil was dried under high vacuum with an auxiliary cold trap in order to obtain an accurate weight of the product, which solidified into a solid mass (23.2 g, 81% yield). This material was dissolved in hexane to give a 1 M solution that was stored in a refrigerator. Upon cold storage, dicyclohexylboron trifluoromethanesulfonate crystallizes from solution, but the crystals easily redissolve when the solution is stirrred at room temperature before use. Reagent solutions prepared in this way gave excellent results in the aldol reaction, p. 116.

15. A stock solution of dicyclohexylboron trifluoromethanesulfonate in hexane is reasonably stable, if stored at ~4°C under nitrogen. It may be purified by storing the hexane solution at -20°C, collecting the crystals that form, and redissolving this material in a calculated amount of dry hexane to give a 1 M solution.

Waste Disposal Information

All toxic materials were disposed of in accordance with "Prudent Practices in the Laboratory;" National Academy Press; Washington, DC, 1995.

3. Discussion

Dialkylboron trifluoromethanesulfonates (triflates) are particularly useful reagents for the preparation of boron enolates from carbonyl compounds, including ketones, thioesters and acyloxazolidinones.[4] Recently, the combination of dicylohexylboron trifluoromethanesulfonate and triethylamine was found to effect the enolization of carboxylic esters.[5] The boron-mediated asymmetric aldol reaction of carboxylic esters is particularly useful for the construction of anti β-hydroxy-α-methyl carbonyl units.[6] The present procedure is a slight modification of that reported by Brown, et al.[2]

1. Venture Laboratory, Kyoto Institute of Technology, Matsugasaki, Kyoto, 606-8585, Japan.

2. (a) Brown, H. C.; Kramer, G. W.; Levy, A. B.; Midland, M. M. "Organic Syntheses via Boranes"; Wiley-Interscience: New York, 1975, p 28; (b) Brown, H. C.; Ganesan, K.; Dhar, R. K. *J. Org. Chem.* **1993**, *58*, 147.

3. Soderquist, J. A.; Negron, A. *Org.Synth., Col. Vol. IX* **1998**, 95.

4. See, eg. (a) Kim, B. M.; Williams, S. F.; Masamune, S. Comprehensive Organic Synthesis; Trost, B. M., Fleming, I., Eds.; Pergamon Press: Oxford, 1991; Vol. 2 (Heathcock, C. H., Ed.), Chapter 1.7, p 239; (b) Cowden, C. J.; Paterson, I. *Org. React.* **1997**, *51*, 1.

5. Abiko, A.; Liu, J.-F.; Masamune, S. *J. Org. Chem.* **1996**, *61*, 2590.

6. Abiko, A.; Liu, J.-F.; Masamune, S. *J. Am. Chem. Soc.* **1997**, *119*, 2586. See the procedure for the anti-selective asymmetric aldol reaction of carboxylic esters, p. 116.

Appendix

Chemical Abstracts Nomenclature (Collective Index Number);

(Registry Number)

Dicyclohexylboron trifluoromethanesulfonate: Methanesulfonic acid, trifluoro-, anhydride with dicyclohexylborinic acid (13); (145412-54-0)

Borane- dimethyl sulfide complex: Methyl sulfide, compd. with borane (1:1); Borane, compd. with thiobis[methane] (1:1); (13292-87-0)

Cyclohexene (8,9); (110-83-8)

Dicyclohexylborane: Borane, dicyclohexyl- (8,9); (1568-65-6)

Trifluoromethanesulfonic acid: HIGHLY CORROSIVE: Methanesulfonic acid, trifluoro- (8,9); (1493-13-6)

2-(N-BENZYL-N-MESITYLENESULFONYL)AMINO-1-PHENYL-1-PROPYL PROPIONATE

[[Benzenesulfonamide, 2,4,6-trimethyl-N-[1-methyl-2-(1-oxopropoxy)-2-phenylethyl]-N-(phenylmethyl)-, [R-(R*,S*)]-]

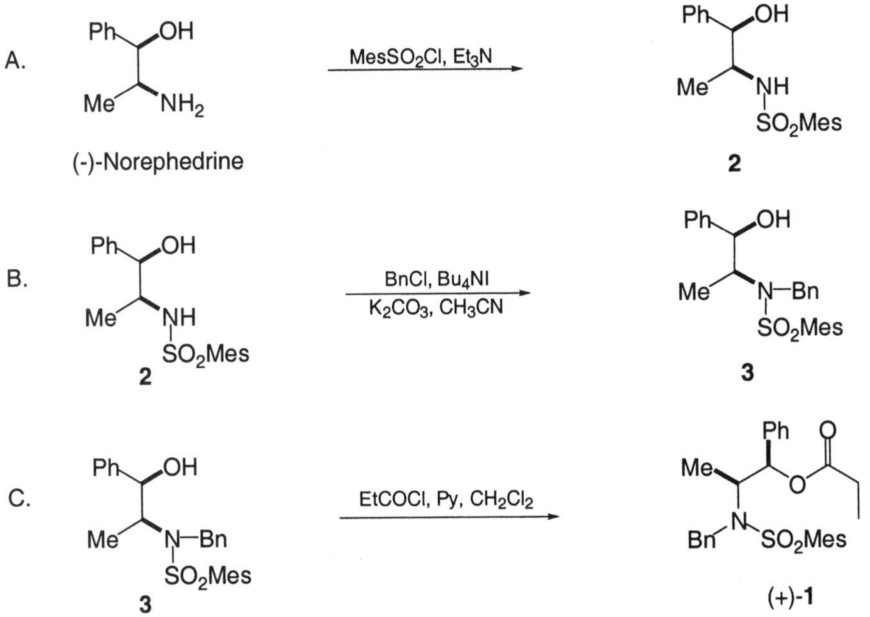

A.

Ph OH
Me NH₂

(-)-Norephedrine

MesSO₂Cl, Et₃N →

Ph OH
Me NH
SO₂Mes

2

B.

Ph OH
Me NH
SO₂Mes

2

BnCl, Bu₄NI
K₂CO₃, CH₃CN →

Ph OH
Me N—Bn
SO₂Mes

3

C.

Ph OH
Me N—Bn
SO₂Mes

3

EtCOCl, Py, CH₂Cl₂ →

Me Ph O
Bn N SO₂Mes O

(+)-1

Submitted by Atsushi Abiko.[1]

Checked by Lloyd J. Simons and William R. Roush.

1. Procedure

A. 2-(N-Mesitylenesulfonyl)amino-1-phenyl-1-propanol, 2. To a stirred solution of (-)-norephedrine (30.2 g, 0.200 mol) (Note 1) and triethylamine (33.4 mL, 0.24 mol) in dichloromethane (400 mL) is added mesitylenesulfonyl chloride (43.8 g, 0.20 mol) (Note 1) at 0°C. The reaction mixture is stirred at 0°C for 2 hr and diluted with diethyl ether (600 mL). The mixture is washed successively with 100 mL each of water, 1 M hydrochloric acid (HCl), water, saturated sodium hydrogen carbonate solution and brine, and dried over anhydrous sodium sulfate. The organic solution is filtered, and the filtrate is concentrated to give an oily residue, which is dissolved in dichloromethane (50 mL). Hexane (100 mL) is added in portions with swirling to the dichloromethane solution to cause crystallization. Additional hexane (300 mL) is added and the crystalline (1R, 2S)-**2** (60.8 g, 91%) is isolated by filtration (Notes 2, 3).

B. *2-(N-Benzyl-N-mesitylenesulfonyl)amino-1-phenyl-1-propanol*, **3**. A mixture of **2** (16.7 g, 50 mmol), benzyl chloride (6.90 mL, 60 mmol) (Note 1), tetrabutylammonium iodide (200 mg) (Note 1) and potassium carbonate (8.4 g, 60 mmol) in acetonitrile (100 mL) is heated under reflux for 17 hr (Note 4). The cooled mixture is filtered and the salt is washed with diethyl ether (100 mL). The combined organic layers are concentrated and the residue is crystallized from dichloromethane (25 mL) and hexane (100 mL) to give **3** (17.0 g, 80%) (Notes 5, 6).

C. *2-(N-Benzyl-N-mesitylenesulfonyl)amino-1-phenyl-1-propyl propionate*, **1**. To a solution of **3** (15.0 g, 35.4 mmol) and pyridine (3.7 mL, 46 mmol) in dichloromethane (200 mL), propionyl chloride (3.8 mL, 44 mmol) (Note 1) is added dropwise at 0°C. The reaction mixture is stirred at room temperature for 13 hr and diluted with diethyl ether (300 mL). The mixture is washed successively with 100 mL each of water, 1 M HCl, water, saturated sodium hydrogen carbonate solution, and brine, and dried with anhydrous sodium sulfate. The filtered organic solution is concentrated to give a crystalline residue, which is triturated with hexane to give **1** (16.8 g, ≥99%) (Note 7).

2. Notes

1. (-)- and (+)-Norephedrine, benzyl chloride, tetrabutylammonium iodide, and propionyl chloride were purchased from Wako Pure Chemical Ltd. Japan or Aldrich Chemical Company, Inc., and used as received. Mesitylenesulfonyl chloride was obtained from Tokyo Kasei Kogyo Ltd. Japan or Aldrich Chemical Company, Inc. and used as received. Dichloromethane and acetonitrile were distilled from calcium hydride (CaH_2) under an inert atmosphere prior to use. Merck 60 silica gel, 0.040-0.063 mm, or Whatman 60 Å 230-400 mesh silica gel was used for column chromatography.

2. The submitter reports that a second crop of pure **2** (6.0 g) could be obtained by concentration and recrystallization of the mother liquors. However, the second crops obtained by the checkers were highly colored and were not sufficiently pure for use in the next step.

3. Sulfonamide **2** exhibited the following physical and spectroscopic properties: mp 121-122°C, TLC (silica gel) R_f = 0.28 in 3:1 hexanes : ethyl acetate; (1R, 2S)-**2**: $[\alpha]_D^{23}$ -12.4 (c 2.12, $CHCl_3$). (*1S, 2R*)-**2**: $[\alpha]_D^{23}$ +12.8 (c 2.12, $CHCl_3$). ^1H NMR (500 MHz, $CDCl_3$) (concentration dependent; 20 mg/mL) δ: 0.87 (d, 3 H, J = 6.8), 2.30 (s, 3 H), 2.52 (1 H, -O<u>H</u>), 2.66 (s, 6 H), 3.53 (m, 1 H), 4.76 (br,1 H, -N<u>H</u>), 4.82 (d,1 H, J = 8.8), 6.96 (s, 2 H), 7.20-7.36 (m, 5 H); (200 mg/mL) δ 0.88 (d, 3 H, J = 6.8), 2.32 (s, 3 H), 2.68 (s, 6 H), 3.08 (1 H, -O<u>H</u>), 3.51 (ddq, 1 H, J = 3.1, 6.8, 9.0), 4.81 (br, 1 H, -N<u>H</u>), 5.23 (br, 1 H), 6.98 (s, 2 H), 7.20-7.36 (m, 5 H). ^{13}C NMR (125 MHz, $CDCl_3$, c = 20 mg/mL) δ: 14.3, 20.9, 22.9, 54.6, 75.6, 125.9, 127.5, 128.3, 132.0, 134.2, 138.9, 140.5, 142.2. IR (thin film from CH_2Cl_2) cm^{-1}: 3504, 3302, 2981, 1064, 1452, 1320, 1156, 1058, 972, 895, 702, 657. HRMS-FAB: Calcd for $C_{18}H_{24}NO_3S$ [M+H]$^+$, 334.1477 *m/z*; Found, 334.1478 *m/z*. Anal. Calcd for $C_{18}H_{23}NO_3S$: C, 64.84; H, 6.95; N, 4.20. Found: C, 64.94; H, 6.98; N, 4.17.

4. The checkers observed that the reaction is not complete according to TLC analysis after the prescribed 17-hr reaction period. However, if the reaction is allowed to proceed for longer reaction times, the dibenzylated product **4** is obtained with correspondingly diminished yields of **3**. When the reaction was terminated after 17 hr, the checkers obtained an 86% yield of recrystallized **3**.

5. The submitter reports that an additional 3.2 g (15%) of **3** was isolated by chromatography (Note 1) of the mother liquor on silica gel (100 g) with 10% ethyl acetate in hexane. The checkers observed that the concentrated mother liquors were not soluble in the chromatography solvent, so they dissolved the viscous oil in dichloromethane (<7 mL) and applied this solution to the column. Once this material was adsorbed, the column was flushed with hexanes (400 mL) to remove the CH_2Cl_2. Elution of the column with 10% ethyl acetate-hexanes then provided additional product **2**. The combined yield of **2** obtained by the checkers via the crystallization and chromatography sequence was 90-93%.

6.   N-Benzylsulfonamide **3** exhibited the following physical and spectroscopic properties: mp 123-124°C; TLC (silica gel) R_f = 0.48 in 3:1 hexanes : ethyl acetate; R_f = 0.15 in 9:1 hexanes : ethyl acetate; (*1R, 2S*)-**3**: $[\alpha]_D^{23}$ -6.3° (c 2.06, CHCl$_3$); (*1S, 2R*)-**3**: $[\alpha]_D^{23}$ +6.4° (c 2.05, CHCl$_3$); ^{1}H NMR (500 MHz, CDCl$_3$) (concentration dependent; 18 mg/mL) δ: 1.03 (d, 3 H, J = 7.0), 2.14 (1 H, -O<u>H</u>), 2.29 (s, 3 H), 2.65 (s, 6 H), 3.82 (dq, 1 H, J = 1.9, 7.0), 4.54 (1 H, A of ABq, J_{AB} = 16.1), 4.77 (1 H, B of ABq, J_{AB} = 16.1), 5.00 (br s, 1 H), 6.93 (s, 2 H), 7.04-7.08 (m, 2 H), 7.10-7.36 (m, 8 H); (218 mg/mL) δ: 1.02 (d, 3 H, J = 7.0), 2.27 (s, 3 H), 2.34 (1 H, -O<u>H</u>), 2.62 (s, 6 H), 3.82 (dq, 1 H, J = 1.9, 7.0), 4.56 (1 H, A of ABq, J_{AB} = 16.1), 4.75 (1 H, B of ABq, J_{AB}

= 16.1 Hz), 4.96 (br s, 1 H), 6.91 (s, 2 H), 7.04-7.08 (m, 2 H), 7.10-7.36 (m, 8 H); ^{13}C

NMR (125 MHz, CDCl$_3$, 18 mg/mL) δ: 9.8, 20.9, 23.0, 49.1, 59.7, 76.6, 125.5, 127.2,

127.4, 127.7, 128.2, 128.6, 132.2, 133.4, 138.6, 140.2, 142.1, 142.6; IR (film from

CH$_2$Cl$_2$) cm^{-1}: 3502, 2981, 1604, 1454, 1314, 1150, 1022, 924, 855, 699, 658; HRMS-

FAB: Calcd for C$_{25}$H$_{30}$NO$_3$S [M+H]$^+$, 424.1946 m/z; Found, 424.1947 m/z; Anal. Calcd

for C$_{25}$H$_{29}$NO$_3$S: C, 70.89; H, 6.90; N, 3.31; Found: C, 70.91; H, 6.95; N, 3.32.

7. The checkers obtained **1** in 96% yield after recrystallization of the crude product

(see Note 8).

8.   (*1R, 2S*)- and (*1S, 2R*)-**1** exist as polymorphic forms. Recrystallization from

hot ethyl acetate (4 mL/g of **1**) and hexane (ethyl acetate : hexane = 1 : 2) afforded higher

melting crystals: mp 124°C, 147-148°C, TLC (silica gel) R$_f$ = 0.52 in 3:1 hexanes : ethyl

acetate; R$_f$ = 0.22 in 9:1 hexanes : ethyl acetate; (*1R, 2S*)-**1** $[\alpha]_D^{23}$+11.1° (c 2.24, CHCl$_3$);

(*1S, 2R*)-**1**: $[\alpha]_D^{23}$ -11.2° (c 2.38, CHCl$_3$); ^{1}H NMR (500 MHz, CDCl$_3$) δ: 1.01 (t, 3 H, J =

7.4), 1.12 (d, 3 H, J = 7.0), 2.14 (m, 2 H), 2.27 (s, 3 H), 2.51 (s, 6 H), 4.04 (dq, 1 H, J =

4.0, 7.0), 4.60 (1 H, A of ABq, J$_{AB}$ = 16.6), 4.72 (1 H, B of ABq, J$_{AB}$ = 16.6), 5.84 (d, 1 H,

J = 3.9), 6.87 (s, 2 H), 6.88-6.96 (m, 2 H), 7.13~7.35 (m, 8 H). ^{13}C NMR (125 MHz

CDCl$_3$) δ: 8.5, 12.3, 20.6, 22.7, 27.1, 47.9, 56.5, 77.7, 125.6, 126.8, 127.1, 127.5, 128.1

(2C), 131.9, 133.2, 138.4, 138.5, 139.9, 142.3, 172.2; IR (film from CH$_2$Cl$_2$) cm^{-1} : 2982,

1747, 1604, 1454, 1381, 1324, 1205, 1153, 1080, 1056, 1020, 859, 764, 730, 699,

659; HRMS-FAB Calcd for C$_{28}$H$_{34}$NO$_4$S [M+H]$^+$, 480.2209 m/z; Found, 480.2186 m/z;

Anal. Calcd for C$_{28}$H$_{33}$NO$_4$S: C, 70.12; H, 6.93; N, 2.92. Found: C, 70.40; H, 6.97; N,

2.90.

Waste Disposal Information

All toxic materials were disposed of in accordance with "Prudent Practices in the

Laboratory"; National Academy Press; Washington, DC, 1995.

3. Discussion

The present procedure is a modification of that originally reported by the submitter and co-workers.[2] This procedure is applicable to a large scale preparation of the title compound in high overall yield (~80%) without purification of the intermediates by chromatography. The title compound is reported to be a useful reagent for *anti*-selective aldol reactions with dicyclohexylboron triflate and triethylamine as enolization reagents.[3]

1. Venture Laboratory, Kyoto Institute of Technology, Matsugasaki, Kyoto, 606-8585, Japan.

2. Abiko, A.; Liu, J.-F.; Masamune, S. *J. Am. Chem. Soc.* **1997**, *119*, 2586.

3. See the procedure describing the anti-selective asymmetric aldol reaction of carboxylic esters, p. 116.

Appendix
Chemical Abstracts Nomenclature (Collective Index Number);
(Registry Number)

2-(N-Benzyl-N-mesitylenesulfonyl)amino-1-phenyl-1-propyl propionate:
Benzenesulfonamide, 2,4,6-trimethyl-N-[1-methyl-2-(1-oxopropoxy)-2-phenylethyl]-N-(phenylmethyl)- (14); [R-(R*,S*)]-, (187324-66-9), [S-(R*,S*)]-, (187324-67-0)

2-(N-Mesitylenesulfonyl)amino-1-phenyl-1-propanol: Benzenesulfonamide, N-(2-hydroxy-1-methyl-2-phenylethyl)-2,4,6-trimethyl-, [S-(R*,S*)]- (14); (187324-62-5)

(1R,2S)-(-)-Norephedrine: Norephedrine (8); Benzeneethanol, α-(1-aminoethyl)-, [R-(R*,S*)]- (9); (492-41-1)

Triethylamine (8); Ethanamine, N,N-diethyl- (8,9); (121-44-8)

Mesitylenesulfonyl chloride: 2-Mesitylenesulfonyl chloride (8); Benzenesulfonyl chloride, 2,4,6-trimethyl- (9); (773-64-8)

2-(N-Benzyl-N-mesitylenesulfonyl)amino-1-phenyl-1-propanol: Benzenesulfonamide, N-(2-hydroxy-1-methyl-2-phenylethyl)-2,4,6-trimethyl-N-(phenylmethyl)- (14); [R-(R*,S*)]-, (187324-63-6), [S-(R*,S*)]-, (187324-64-7)

ANTI-SELECTIVE BORON-MEDIATED ASYMMETRIC ALDOL REACTION OF CARBOXYLIC ESTERS: SYNTHESIS OF (2S, 3R)-2,4-DIMETHYL-1,3-PENTANEDIOL

[1,3-Pentanediol, 2,4-dimethyl-, [S-(R*,S*)-]]

Submitted by Atsushi Abiko.[1]

Checked by Eric M. Flamme and William R. Roush.

1. Procedure

A. (1'R)-Phenyl-(2'S)-[(phenylmethyl)[(2,4,6-trimethylphenyl)sulfonyl]amino]-propyl (3R)-hydroxy-(2R),4-dimethylpentanoate.[2] An oven-dried, 500-mL, round-bottomed flask is charged with (1R, 2S)-(+)-1 (4.80 g, 10 mmol) (Note 1) and dichloromethane (50 mL) (Note 2) under nitrogen. To this solution is added triethylamine (3.40 mL, 24 mmol) (Note 2) via syringe. The solution is cooled to -78°C and a solution of dicyclohexylboron triflate (1.0 M in hexane, 22 mL, 22 mmol) (Note 3) is added dropwise over 20 min. The resulting solution is stirred at -78°C for 30 min. To the -78°C enolate solution is then added isobutyraldehyde (1.08 mL, 12 mmol, freshly distilled) dropwise.

The reaction mixture is stirred for 30 min at -78°C and allowed to warm to room temperature over 1 hr, then quenched by the addition of pH 7 buffer solution (40 mL). The mixture is diluted with methanol (MeOH, 200 ml) and 30% hydrogen peroxide (20 mL) is added carefully (Note 4). The mixture is stirred vigorously overnight and then concentrated on a rotary evaporator. The residue is partitioned between water (100 mL) and dichloromethane (200 mL). The aqueous layer is extracted with dichloromethane (150 mL x 2). The combined organic extracts are washed with water (100 mL x 3) and dried with anhydrous sodium sulfate. The solids are removed by flitration, and the organic layer is concentrated. The residue (Notes 5, 6) is crystallized from 80% hexane-20% ethyl acetate (EtOAc, 80 mL) to afford (+)-2 (4.1 g, 74%) with a diastereomeric purity of 24 : 1 (Note 7). The mother liquors are concentrated using a rotary evaporator, and the residue is diluted with mesitylene (50 mL). This mixture is distilled at ~60°C (0.1 mm) to remove cyclohexanol. The resulting material is purified by chromatography over silica gel (30 g) (Note 2) using a 9 : 1 : 1 mixture of hexane, ethyl acetate and dichloromethane (Note 8) to give isomerically pure (+)-2 (0.8 g, 14%) (Note 9). If desired, the material from the first crystallization may also b e purified by silica gel chromatography to give diastereomerically pure product (Note 10).

   B. *(2S, 3R)-2,4-Dimethyl-1,3-pentanediol 3*. To a stirred solution of (+)-2 (2.75 g, 5 mmol, 96 : 4 isomeric purity) in tetrahydrofuran (THF) (50 mL) is added lithium aluminum hydride (0.19 g, 5 mmol) at 0°C. The reaction mixture is stirred at room temperature for 1 hr and quenched by the careful addition of sodium sulfate decahydrate (5 g). The mixture is stirred vigorously for 30 min and filtered. The filtrate is concentrated, dissolved in 75 mL of a 1:1 mixture of hexane and dichloromethane. This solution is dried over sodium sulfate, filtered and concentrated under reduced pressure. Trituration of the resulting oil with hexane (50 ml) results in the precipitation of auxiliary alcohol 4 (1.6-1.8 g) which is recovered by filtration (Note 11). The residue is separated by chromatography over silica gel (40 g) (Note 2) with hexane and ethyl acetate (3:1-1:1) to afford additional 4 (0.2-0.4 g, Note 12) and 3 (0.60 g, 92%) (Notes 13, 14).

117

2. Notes

1. (1R,2S)-(+)-**1** was prepared according to the accompanying procedure (Abiko, A. *Org. Synth.* **2002**, *79*, 109).

2. Dichloromethane and triethylamine were distilled from calcium hydride before use. Silica gel 60, 0.040-0.063 mm (Merck) was used for column choromatography. Lithium aluminum hydride (95%, powder) was obtained from Aldrich Chemical Co., Inc.

3. A stock solution (1 M) of dicyclohexylboron trifluromethanesulfonate was prepared according to the accompanying procedure (Abiko, A. *Org. Synth.* **2002**, *79*, 103). Two equivalents of the boron triflate are necessary for complete enolization of the ester. When one equivalent is used, the enolization proceeds only to 50% conversion.

4. Addition of hydrogen peroxide may cause a vigorous exothermic reaction.

5. The reaction may be monitored by TLC using a solvent mixture of 90% hexanes and 10% ethyl acetate. Starting material **1** has an $R_f = 0.37$ and the product **2** has an $R_f = 0.25$. Both spots were visualized by UV detection. It is possible to visualize the minor diastereoisomer by using a dilute TLC sample and a solvent system of 85% hexanes and 15% ethyl acetate. Under these conditions product **2** has an $R_f = 0.32$ and the minor aldol diastereoisomer has an $R_f = 0.27$.

6. HPLC analysis [21 mm Dynamax-60A column (Si 83-111-C), 78% hexanes-22% ethyl acetate, 10 ml/min flow rate, UV detection; retention time for **1** = 14 min; retention time for **2** = 16.5 min; retention time for minor aldol diastereoisomer = 17.9 min] of the crude reaction product showed that the residue contained 10-15% of the starting ester and that the two anti aldol products were obtained in 96 : 4 ratio. The submitter found the product ratio to be 97 : 3 by HPLC analysis, with less than 3% of the starting ester remaining.

7. Attempts to further purify this material by additional recrystallizations were unsuccessful. The submitter indicated that diastereomerically pure **2** could be obtained by additional recrystallizations from 85 : 15 hexanes-ethyl acetate, but the checkers were unable to achieve this result. However, the two aldol diastereomers can be separated chromatographically, as described in the body of the procedure.

8. The submitter performed this chromatography step using a 5 : 1 mixture of hexane and ethyl acetate. However, the checkers found that the mother liquors did not dissolve in this solvent mixture, and added dichloromethane to the chromatography mixture to solve this problem.

9. The physical and spectral data of (+)-2 are as follows: mp 142-142.5°C, $[\alpha]_D^{23}$ 19.7° (c 2.05, CHCl$_3$); ^{1}H NMR (CDCl$_3$) δ: 0.90 (d, 3H, J = 6.7), 0.95 (d, 3H, J = 6.8), 1.10 (d, 3H, J = 7.2), 1.17 (d, 3H, J = 7.0), 1.73, (m, 1H), 2.28 (s, 3H), 2.37 (br s, 1H, OH), 2.49 (s, 6H), 2.62 (dq, 1H, J = 7.1, 7.2), 3.41 (br, 1H), 4.11 (dq, 1H, J = 4.4, 7.0), 4.55 (1H, A of ABq, J_{AB} = 16.5), 4.79 (1H, B of ABq, J_{AB} = 16.5), 5.82 (d, 1H, J = 4.4), 6.82-6.86 (m, 2H), 6.87 (s, 2H), 7.12-7.33 (m, 8H); ^{13}C NMR (CDCl$_3$) δ: 13.4, 14.2, 15.5, 19.8, 20.7, 22.8, 30.0, 42.9, 48.1, 56.7, 77.6, 78.1, 125.8, 127.0, 127.6, 127.8, 128.2, 128.3, 132.0, 133.3, 138.1, 138.5, 140.1, 142.4, 174.8. Anal. Calcd for C$_{32}$H$_{41}$NO$_5$S: C, 69.66; H, 7.49; N, 2.54. Found: C, 69.84; H, 7.62; N, 2.53.

10. In one run (5.0-mmol scale), the checkers obtained a crude product that contained ca. 40% of recovered 1. Attempts to purify the aldol product from this mixture by using the described crystallization procedure was unsuccessful. Accordingly, the crude product (2.36 g) was purified by flash chromatography on 170 g of silica gel using 1000 ml of a 9 : 1 : 1 mixture of hexane, ethyl acetate and methylene chloride. This provided 0.72 g of starting ester 1, 1.39 g of pure aldol 2, and 0.259 g of mixed fractions containing 2 and the minor aldol diastereoisomer (4 : 1 by ^{1}H NMR analysis).

11. If the auxiliary does not precipitate during the trituration step, the solution is not sufficiently dry or may contain too much dichloromethane. Under these circumstances, the solution should be concentrated, redried with sodium sulfate, and the trituration procedure repeated. The checkers also found that trituration is best performed by spinning the flask on a rotary evaporator (at atmospheric pressure) for several minutes. Swirling the flask by hand was not always successful.

12. The checkers found that small amounts of ester 2 (2-7%) remained unreacted, even when 1.1 equiv of lithium aluminum hydride was used for the reduction. Progress of the reduction can be monitored by TLC (70% hexane, 30% ethyl acetate): 2, R_f = 0.75; 3,

119

Rf = 0.2; **4**, R$_f$ = 0.75. Because **2** and **4** co-elute under the chromatography conditions, auxiliary **4** recovered by chromatography is not pure.

13. The submitter and checkers have successfully performed this procedure on double the reaction scale [10 mmol of (+)-**2**], and obtained the product in 92% yield.

14. The spectroscopic properties of product (2S, 3R)-**3** are as follows: $[\alpha]_D^{23}$ 19.6° (c 0.57, CHCl$_3$), lit.[2] 19.6° (c 0.75, CHCl$_3$); ^{1}H NMR (CDCl$_3$) δ: 0.78 (d, 3 H, J = 7), 0.81 (d, 3 H, J = 6.6), 0.87 (d, 3 H, J = 7.0), 1.73 (m, 2H), 3.24(dd, 1 H, J = 3.4, 8.0), 3.53 (1 H, B of ABX, J$_{AB}$ =10.8, J$_{BX}$ = 7), 3.65 (1 H, A of ABX, J$_{AB}$ = 10.8, J$_{AX}$ = 3.7), 3.70 (br s, 2 H, OH); ^{13}C NMR (CDCl$_3$) δ: 14.0, 15.2, 20.0, 30.5, 37.2, 68.0, 81.8; Anal. Calcd for C$_7$H$_{16}$O$_2$: C, 63.60; H, 12.20. Found: C, 63.42; H, 12.01.

Waste Disposal Information

All toxic materials were disposed of in accordance with "Prudent Practices in the Laboratory"; National Academy Press; Washington, DC, 1995.

3. Discussion

Several methods for the anti-selective, asymmetric aldol reaction[3] recorded in the literature include (i) the use of boron, titanium, or tin(II) enolate carrying chiral ligands,[4] (ii) Lewis acid-catalyzed aldol reactions of a metal enolate of chiral carbonyl compounds,[5] and (iii) the use of the metal enolate derived from a chiral carbonyl compound.[6] Although many of these methods provide anti-aldols with high enantioselectivities, these methods are not as convenient or widely applicable as the method reported here, because of problems associated with the availability of reagents, the generality of reactions, or the required reaction conditions.

The present procedure is based on the original report by the author and co-workers,[7] and utilizes the characteristic features of the boron-mediated aldol reaction of carboxylic esters, represented by the ability to produce either anti- or syn-aldols under the specified reaction conditions.[8] The reliability and practicality of the boron- mediated aldol reaction have been demonstrated by many examples.[9] Both enantiomers of the chiral auxiliary alcohol in this procedure are prepared from readily available (-)- or (+)-norephedrine in three easy steps in high overall yield.[10] The auxiliary alcohol could be recovered in nearly quantitative yield (and reused) with the transformation of the aldol products to chiral diols or other derivatives. The stoichiometry of the boron triflate required for the enolization of carboxylic esters was determined empirically. The present procedure is applicable to a wide range of aldehydes with high selectivity (both syn:anti and diastereofacial selectivity of anti isomer); see the following Table.[7] An application to a natural product synthesis has been reported[11]

Table Boron-Mediated Asymmetric Aldol Reactions

Aldehyde, R-	yield (%)[a]	ds for anti[a]	product	mp (°C); [α]D in CHCl3
Et-	90	96:4	**2a**	128-129; 16.2° (c 1.52)
c-Hex-	91	95:5	**2b**	glass; 21.7° (c 2.40)
tert-Bu-	96	99:1	**2c**	185-186; 22.1° (c 1.70)
Ph-	93	95:5	**2d**	glass; 18.9° (c 1.70)
(E)-MeCH=CH-	96	98:2	**2e**	glass; 24.8° (c 1.25)
CH₂=C(Me)-	97	96:4	**2f**	97-98; 19.7° (c 1.32)
BnOCH₂CH₂-	94	95:5	**2g**	glass; 16.2° (c 1.30)
BnOCH₂C(Me)₂-	98	96:4	**2h**	109-110; 16.2° (c 1.30)

[a] Yield and isomer ratio by HPLC. Syn isomers <2%.

121

1. Venture Laboratory, Kyoto Institute of Technology, Matsugasaki, Kyoto, 606-8585, Japan.

2. Masamune, S.; Sato, T.; Kim, B. M.; Wollmann, T. A. *J. Am. Chem. Soc.* **1986**, *108*, 8279.

3. For instance, see: (a) Kim, B.-M.; Williams, S. F.; Masamune, S. "Comprehensive Organic Synthesis"; Trost, B. M., Fleming, I., Eds.; Pergamon Press: Oxford, **1991**; Vol. 2 (Heathcock, C. H., Ed.), Chapter 1.7, p 239. (b) Heathcock, C. H. "Modern Synthetic Methods"; Scheffold, R., Ed.; VCH: New York, **1992**; p 1; (c) Braun, M.; Sacha, H. *J. Prakt. Chem.* **1993**, *335*, 653.

4. For Sn enolates, see: (a) Narasaka, K.; Miwa, T. *Chem. Lett.* **1985**, 1217. For Ti enolates, see: (b) Duthaler, R. O.; Herold, P.; Wyler-Helfer, S.; Riediker, M. *Helv. Chim. Acta* **1990**, *73*, 659. For B enolates, see: (c) Meyers, A. I.; Yamamoto, Y. *Tetrahedron* **1984**, *40*, 2309; (d) Masamune, S.; Sato, T.; Kim, B.-M.; Wollman, T. A. *J. Am. Chem. Soc.* **1986**, *108*, 8279; (e) Reetz, M. T.; Rivadeneira, E.; Niemeyer, C. *Tetrahedron Lett.* **1990**, *27*, 3863; (f) Corey, E. J.; Kim, S. S. *J. Am. Chem. Soc.* **1990**, *112*, 4976; (g) Gennari, C.; Hewkin, C. T.; Molinari, F.; Bernardi, A.; Comotti, A.; Goodman, J. M.; Paterson, I. *J. Org. Chem.* **1992**, *57*, 5173; (h) Gennari, C.; Moresca, D.; Vieth, S.; Vulpetti, A. *Angew. Chem., Int. Ed. Engl.* **1993**, *32*, 1618.

5. For Si enolates, see: (a) Helmchen, G.; Leikauf, U.; Taufer-Knöpfel, I. *Angew. Chem., Int. Ed. Engl.* **1985**, *24*, 874; (b) Gennari, C.; Bernardi, A.; Colombo, L.; Scolastico, C. *J. Am. Chem. Soc.* **1985**, *107*, 5812; (c) Oppolzer, W.; Marco-Contelles, J. *Helv. Chim. Acta* **1986**, *69*, 1699; (d) Oppolzer, W.; Starkemann, C.; Rodriguez, I.; Bernardinelli, G. *Tetrahedron Lett.* **1991**, *32*, 61; (e) Oppolzer, W.; Lienard, P. *Tetrahedron Lett.* **1993**, *34*, 4321. For B enolates, see: (f) Danda, H.; Hansen, M. M.; Heathcock, C. H. *J. Org. Chem.* **1990**, *55*, 173; (g) Walker, M. A.; Heathcock, C. H. *J. Org. Chem.* **1991**, *56*, 5747; (h) Wang, Y.-

C.; Hung, A.-W.; Chang, C.-S.; Yan, T.-H. *J. Org. Chem.* **1996**, *61*, 2038. For Ti enolates, see: (i) Ghosh, A. K.; Ohnishi, M. *J. Am. Chem. Soc.* **1996**, *118*, 2527.

6. (a) Davies, S. G.; Dordor-Hedgecock, I. M.; Warner, P. *Tetrahedron Lett.* **1985**, *26*, 2125; (b) Myers, A. G.; Widdowson, K. L. *J. Am. Chem. Soc.* **1990**, *112*, 9672; (c) Van Draanen, N. A.; Arseniyadis, S.; Crimmins, M. T.; Heathcock, C. H. *J. Org. Chem.* **1991**, *56*, 2499; (d) Braun, M.; Sacha, H. *Angew. Chem., Int. Ed. Engl.* **1991**, *30*, 1318; (e) Paterson, I.; Wallace, D. J.; Velazquez, S. M. *Tetrahedron Lett.* **1994**, *35*, 9083; (f) Kurosu, M.; Lorca, M. *J. Org. Chem.* **2001**, *66*, 1205

7. Abiko, A.; Liu, J.-F.; Masamune, S. *J. Am. Chem. Soc.* **1997**, *119*, 2586.

8. Abiko, A.; Liu, J.-F.; Masamune, S. *J. Org. Chem.* **1996**, *61*, 2590. A syn selective boron-mediated asymmetric aldol reaction of a related chiral ester was reported: Liu, J.-F.; Abiko, A.; Pei, Z.-g.; Buske, D. C.; Masamune, S. *Tetrahedron Lett.* **1998**, *39*, 1873.

9. Cowden, C. J.; Paterson, I. *Org. React.* **1997**, *51*, 1.

10. For the preparation of **1**, see: Abiko, A. *Org. Synth.* **2002**, *79*, 109.

11. Yoshimitsu, T.; Song, J. J.; Wang, G.-Q.; Masamune, S. *J. Org. Chem.* **1997**, *62*, 8978.

Appendix

Chemical Abstracts Nomenclature (Collective Index Number); (Registry Number)

(1'R)-Phenyl-(2'S)-[(phenylmethyl)[(2,4,6-trimethylphenyl)sulfonyl]amino]propyl (3R)-hydroxy-(2R), 4-dimethylpentanoate: Pentanoic acid, 3-hydroxy-2,4-dimethyl-1-phenyl-2-

[(phenylmethyl)[(2,4,6-trimethylphenyl)sulfonyl]amino]propyl ester, [2R-[1(1R*,

2S*)2R*,3R*]]- (14); (187324-70-5)

2(N-Benzyl-N-mesitylenesulfonyl)amino-1-phenyl-1-propyl propionate:

Benzenesulfonamide, 2,4,6-trimethyl-N-[1-methyl-2-(1-oxopropoxy)-2-phenylethyl]-N-

(phenylmethyl)-, [R-(R*,S*)]- (14); (187324-66-9)

Triethylamine (8); Ethanamine, N,N-diethyl- (9); (121-44-8)

Dicyclohexylboron trifluoromethanesulfonate: Methanesulfonic acid, trifluoro-, anhydride with

dicyclohexylborinic acid (13); (145412-54-0)

Isobutyraldehyde (8); Propanal, 2-methyl- (9); (78-84-2)

Hydrogen peroxide (8,9); (7722-84-1)

(2S,3R)-2,4-Dimethyl-1,3-pentanediol: 1,3-Pentanediol, 2,4-dimethyl-, [S-(R*,S*)]- (12);

(129262-73-3)

Lithium aluminum hydride: Aluminate (1-), tetrahydro-, lithium (8); Aluminate (1-),

tetrahydro-, lithium, (T-4) (9); (16853-85-3)

CAMPHORQUINONE AND CAMPHORQUINONE MONOXIME

[Bicyclo[2.2.1]heptane-2,3-dione, 1,7,7-trimethyl-, (1R)- and
Bicyclo[2.2.1]heptan-2-one, 1,7,7-trimethyl-3-oxime, (1R)-]

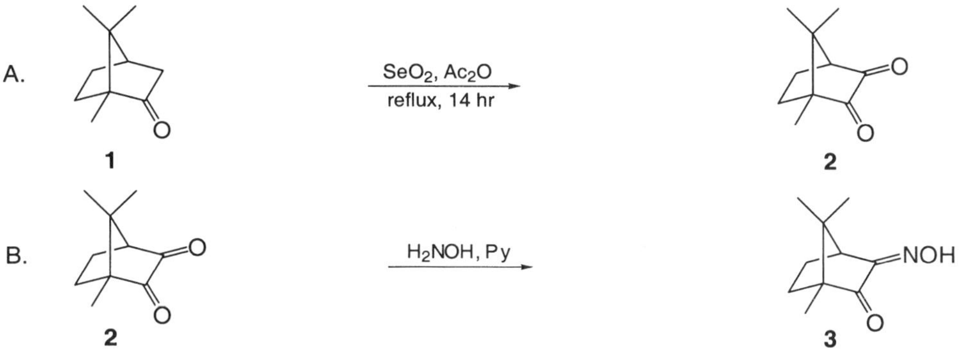

A.

SeO$_2$, Ac$_2$O
reflux, 14 hr

1

2

B.

H$_2$NOH, Py

2

3

Submitted by James D. White, Duncan J. Wardrop and Kurt F. Sundermann.[1]
Checked by Kenji Koga, Kei Manabe, Christopher E. Neipp and Stephen F. Martin.

1. Procedure

Caution! Parts A and B should be carried out in a well ventilated hood since toxic selenium dioxide is used.

A. *(1R,4S)-(-)-Camphorquinone* (Note 1). A 125-mL, three-necked, round-bottomed flask, equipped for mechanical stirring and outfitted with a reflux condenser, is charged with 20.0 g (0.13 mol) of (+)-camphor (**1**) (Note 2), 8.0 g (0.07 mol) of selenium dioxide (*Caution! Selenium dioxide is toxic*) (Note 3), and 14.0 mL of reagent grade acetic anhydride (Note 4). The green mixture is stirred at reflux for 1 hr, cooled to ambient temperature, and an additional 8.0 g (0.07 mol) of selenium dioxide is added. The mixture is again heated to reflux, and two further batches of 8.0 g (0.07 mol) of selenium dioxide

125

are added at 2.5-hr and 6-hr intervals. After the reaction is heated at reflux for an additional 8 hr, during which time precipitation of selenium metal is observed, it is cooled to ambient temperature and transferred to a 125-mL beaker with the aid of 50 mL of ethyl acetate. The black precipitate is removed by filtration, and the filtrate is diluted with 100 mL of toluene (Note 5). Concentration of the filtrate by rotary evaporation gives crude **2** as an orange solid. This is dissolved in 200 mL of ethyl acetate, and the solution is filtered by vacuum filtration through Celite. The filtrate is transferred to a 1-L separatory funnel and is washed successively with 200 mL of 10% aqueous sodium hydroxide solution and 100 mL of saturated aqueous sodium chloride solution. The organic solution is dried over magnesium sulfate, filtered, and concentrated by rotary evaporation to afford 20.89 g of crude (1R,4S)-(-)-camphorquinone (**2**) as yellow crystals, mp 198-199°C (Notes 6 and 7). This material is suitable for the next reaction without purification.

B. *(1R,4S)-(-)-Camphorquinone monoxime.* A 500-mL, single-necked, round-bottomed flask equipped with a magnetic stirring bar is charged with 10.0 g (0.060 mol) of crude (1R)-(-)-camphorquinone (**2**), 240 mL of ethanol, 40 mL of pyridine (Note 8) and 5.44 g (0.078 mol) of hydroxylamine hydrochloride ($NH_2OH \cdot HCl$) (Note 9). The solution is stirred for 20 min, and the ethanol is removed by rotary evaporation at 40°C. The resulting oil is diluted with 100 mL of hexane and 100 mL of ethyl acetate, and the solution is transferred to a 1-L separatory funnel. The organic phase is separated and washed successively with 125 mL of 5% hydrochloric acid solution, 300 mL of water, and 300 mL of saturated aqueous sodium chloride solution. The organic solution is dried over magnesium sulfate, filtered, and concentrated by rotary evaporation. The residue is taken up in 65 mL of heptane and heated to reflux. After reflux is maintained for 2 min, the mixture is allowed to cool to room temperature. The solid is collected by vacuum filtration, and the filter cake is dried under high vacuum to provide 8.63 g (79%) of (1R)-(-)-camphorquinone oxime (**3**) as an off-white solid (Note 10). This is a mixture of syn- and anti-oxime isomers, mp 148-151°C $[\alpha]_D^{22} +184°$ (CH_2Cl_2, *c* 3.5) (Note 11).

2. Notes

1. Camphorquinone is available in racemic and both enantiomeric forms from Aldrich Chemical Company, Inc.

2. (+)-Camphor was purchased from Aldrich Chemical Company, Inc. and was used without purification.

3. Selenium dioxide was purchased from Aldrich Chemical Company, Inc.

4. Reagent grade acetic anhydride was purchased from Fischer Scientific.

5. Addition of toluene at this stage aids the removal of traces of acetic acid.

6. The crude product (2) is of sufficient purity for the next step even if trace amounts of acetic acid and selenium dioxide remain. The material can be purified by crystallization from a mixture of hexane and 2-propanol to provide a product with mp 198-199°C (lit. mp 199°C)[3] and $[\alpha]_D^{22}$ -103° (toluene, c 2.0).

7. The spectral properties of (1R,4S)-(-)-camphorquinone (2) are as follows: ^{1}H NMR (300 MHz, CDCl$_3$) δ: 0.92 (s, 3 H), 1.03 (s, 3 H), 1.08 (s, 3 H), 1.52-1.68 (m, 2 H), 1.79-1.94 (m, 1 H), 2.06-2.22 (m, 1 H), 2.62 (d, 1 H, J = 5.4); ^{13}C NMR (75 MHz, CDCl$_3$) δ: 8.9, 17.6, 21.3, 22.4, 30.1, 42.8, 58.1, 58.8, 203.0, 205.0; IR (film) cm^{-1}: 1767, 1746, 994

8. Reagent grade pyridine was purchased from Fischer Scientific Company.

9. Analytical reagent grade hydroxylamine hydrochloride was purchased from Mallinckrodt Inc.

10. The camphorquinone monoxime 3 is not completely soluble in 65 mL of boiling heptane, but it is nevertheless obtained in pure form after this step.

11. The spectral properties of syn- and anti-(1R,4S)-(-)-camphorquinone monoxime (3) are as follows: ^{1}H NMR (300 MHz, CDCl$_3$) δ: 0.84 (s, 2.6 H), 0.88 (s, 0.4 H), 0.96 (s, 3 H), 0.97 (s, 0.4 H), 0.98 (s, 2.6 H), 1.47-2.12 (comp, 5 H), 2.68 (d, 0.1 H, J = 4.1), 3.23 (d, 0.9 H, J = 4.6); ^{13}C NMR (75 MHz, CDCl$_3$) δ (major isomer): 8.9, 17.6, 20.6, 23.7,

30.6, 44.8, 46.6, 58.5, 159.6, 204.3; IR (CHCl$_3$) cm^{-1}: 3262, 2963, 2878, 1748, 1654; mass spectrum (CI) m/z 182.1179 [C$_{10}$H$_{15}$NO$_2$ (M+1) requires 182.1181], 182 (base), 136.

Waste Disposal Information

All toxic materials were disposed of in accordance with "Prudent Practices in the Laboratory"; National Academy Press; Washington, DC, 1995.

3. Discussion

(-)-Camphorquinone (2) is most conveniently prepared from (1R,4R)-(+)-camphor (1) by the method of Rupe,[2] and is converted to a mixture of syn and anti monoximes 3 by the method of Cherry, et al.[3] The oxime 3 has also been obtained by nitrosation of camphor.[4]

1. Department of Chemistry, Oregon State University, Corvallis, OR 97331-4003.
2. Rupe, H.; Tommasi di Vignano, A. *Helv. Chim. Acta* **1937**, *20*, 1078.
3. Cherry, P. C.; Cottrell, W. R. T.; Meakins, G. D.; Richards, E. E *J. Chem. Soc. (C)* **1968**, 459.
4. Forster, M. O.; Rao, K. S. N. *J. Chem. Soc.* **1926**, 2670.

Appendix

Chemical Abstracts Nomenclature (Collective Index Number); (Registry Number)

(1R,4S)-(-)-Camphorquinone: Bicyclo[2.2.1]heptane-2,3-dione, 1,7,7-trimethyl-, (1R)-; (10334-26-6)

(1R,4S)-(-)-Camphorquinone monoxime: 2,3-Bornanedione, 3-oxime (9); Bicyclo[2.2.1]heptan-2-one, 1,7,7-trimethyl-, 3-oxime, (1R)- (12); (663-17-2)

(+)-Camphor: Camphor, (1R,4R)-(+)- (8);Bicyclo[2.2.1]heptan-2-one, 1,7,7-trimethyl-, (1R)-; (464-49-3)

Selenium dioxide: Selenium oxide (8,9); (7446-08-4)

Acetic anhydride (8); Acetic acid, anhydride (9); (108-24-7)

Hydroxylamine hydrochloride (8); Hydroxylamine, hydrochloride (9); (5470-11-1)

(2S)-(-)-3-exo-(DIMETHYLAMINO)ISOBORNEOL [(2S)-(-)-DAIB]

(Bicyclo[2.2.1]heptan-2-ol, 3-(dimethylamino)-1,7,7-trimethyl-, [1R-(exo,exo)])

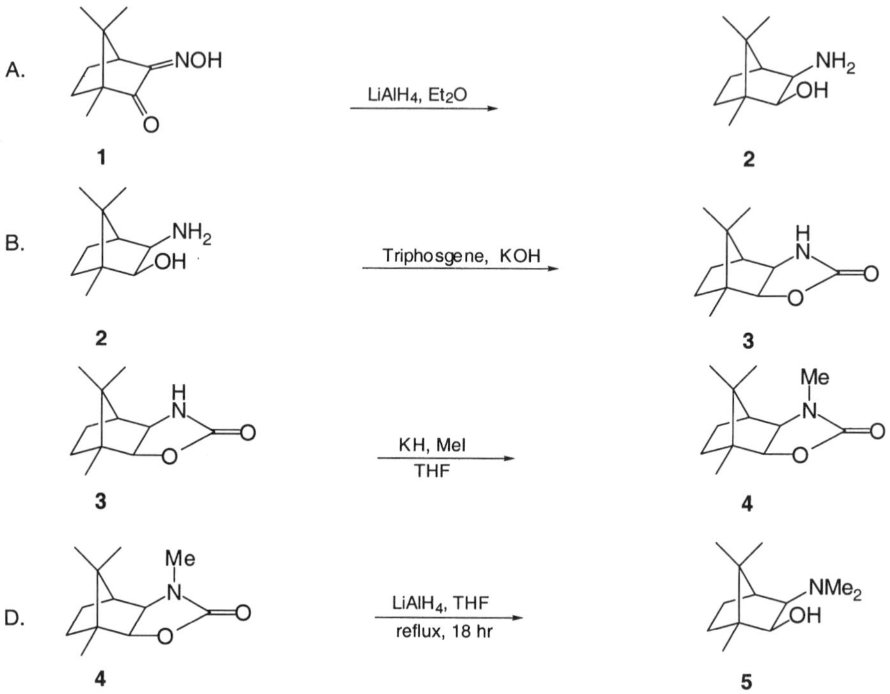

A.
NOH
O
1
LiAlH4, Et2O
NH2
OH
2
B.
NH2
OH
2
Triphosgene, KOH
H
N
O
O
3
H
N
O
O
3
KH, MeI
THF
Me
N
O
O
4
D.
Me
N
O
O
4
LiAlH4, THF
reflux, 18 hr
NMe2
OH
5

Submitted by James D. White, Duncan J. Wardrop, and Kurt F. Sundermann.[1]

Checked by Christopher E. Neipp and Stephen F. Martin.

1. Procedure

A. (2S)-(-)-3-exo-Aminoisoborneol, **2**. All glassware for this step must be oven-dried prior to use. A 2-L, three-necked, round-bottomed flask equipped with an efficient mechanical stirrer attached to a Teflon stirring blade, a thermometer, and a 1-L pressure-equalizing addition funnel connected to an argon inlet is flushed with argon and charged with

475 mL of dry diethyl ether (Note 1) and 5.12 g (0.14 mol) of lithium aluminum hydride (Note 2). The addition funnel is charged with a solution of 8.46 g (0.047 mol) of (1R,4S)-(-)-camphorquinone monoxime (1) (Note 3) in 350 mL of dry ether. The reaction mixture is cooled to 0°C in an ice-water bath with stirring, and the solution of 1 is added slowly to the reaction mixture over 30 min (Note 4). After the addition is complete, the addition funnel is replaced with a reflux condenser connected to the argon inlet, and the reaction flask is removed from the cooling bath. The reaction mixture is stirred and heated at reflux for 1.5 hr and then cooled to 0°C in an ice-water bath. The reaction is quenched very carefully by dropwise addition of 75 mL of saturated aqueous sodium sulfate solution (Notes 5 and 6). The resulting white granular precipitate or slurry is removed by vacuum filtration through Celite, and the filter cake is washed with three portions of 75 mL of chloroform. The combined filtrate and washings are dried over sodium sulfate, filtered, and concentrated by rotary evaporation at 40°C to give 8.07 g of crude (2S)-(-)-3-exo-aminoisoborneol (2) as an off-white solid (Note 7). This material is suitable for the next reaction without purification.

B.   (1R,2S,6R,7S)-1,10,10-Trimethyl-4-oxo-5-aza-3-oxatricyclo[5.2.1.0]decane, 3.. A 500-mL, single-necked, round-bottomed flask equipped with a 250-mL pressure-equalizing dropping funnel and a magnetic stirring bar is charged with 8.06 g (0.048 mmol) of crude (2S)-(-)-3-exo-aminoisoborneol (2), 70 mL of toluene, and 55 mL (0.12 mol) of 12.5% aqueous potassium hydroxide solution. The reaction flask is cooled to 0°C in an ice-water bath, the dropping funnel is charged with a solution of 50 mL (0.095 mol) of triphosgene (1.9 M in toluene) (Note 8), and the triphosgene solution is slowly added over 40 min with vigorous stirring to the two-phase reaction mixture. After the reaction is complete, the reaction mixture is stirred at 0°C for 1 hr and diluted with 130 mL of ethyl acetate. The mixture is transferred to a 1-L separatory funnel, and the layers are separated. The organic layer is washed successively with 150 mL of saturated aqueous sodium bicarbonate solution (Note 9) and 150 mL of saturated sodium chloride solution. The organic layer is dried over magnesium sulfate, filtered, and concentrated by rotary evaporation to give 8.07 g of crude (1R,2S,6R,7S)-1,10,10-trimethyl-4-oxo-5-aza-3-oxatricyclo[5.2.1.0]decane (3) as a waxy pale yellow solid that is suitable for further reaction without purification (Notes 10 and 11).

C. *N-Methyl-(1R,2S,6R,7S)-1,10,10-trimethyl-4-oxo-5-aza-3-oxatricyclo[5.2.1.0]-decane*, 4. All glassware for this step must be oven-dried. A 500-mL, three-necked, round-bottomed flask equipped with a magnetic stirring bar, an argon inlet, a 125-mL pressure-equalizing addition funnel, and a rubber septum is flushed with argon and charged with 9.47 g (0.083 mol) of a 35% dispersion of potassium hydride (KH) in mineral oil (Note 12). The mineral oil is removed by introducing 85 mL of hexane with a syringe, gently stirring the mixture, allowing the potassium hydride to settle, and then withdrawing the liquid by syringe. The KH is washed twice using this procedure. The flask is further charged with 300 mL of dry tetrahydrofuran (THF), Note 13), and the dropping funnel is charged by syringe with a solution of 8.06 g (0.041 mol) of crude (1R,2S,3R,4S)-3-exo-amino-2-exo-hydroxy-1,7,7-trimethylbicyclo[2.2.1]heptane (3) in 90 mL of dry THF. The magnetically stirred suspension of potassium hydride is cooled to 0°C, and the solution of 3 is added dropwise over a 10-min period. After the addition is complete, the mixture is stirred at 0°C for 20 min, and 13 mL (0.21 mol) of methyl iodide (Note 14) is introduced dropwise via syringe. The cooling bath is removed, and the reaction mixture is stirred at room temperature for 10 hr. The reaction is quenched by cautious addition of 15 mL of water, and the resulting mixture is transferred to a 1-L separatory funnel containing 125 mL of water. The layers are separated, and the aqueous layer is extracted three times with 100 mL of ethyl acetate. The combined organic extracts are washed with 300 mL of saturated sodium chloride solution, dried over magnesium sulfate, and concentrated by rotary evaporation to give crude 4. Recrystallization of this material from 15 mL of cyclohexane provides brown crystals, which are dissolved in 15 mL ethyl acetate and decolorized with Norit SA3 (100 mesh) activated carbon at reflux. This solution is cooled to room temperature, and the Norit is removed by vacuum filtration through Celite. The filtrate is concentrated by rotary evaporation, and the resulting pale yellow crystals are recrystallized twice from 5 mL of cyclohexane to provide 2.77 g (32%) of N-methyl-(1R,2S,6R,7S)-1,10,10-trimethyl-4-oxo-5-aza-3-oxatricyclo[5.2.1.0]decane (**4**) as colorless crystals, mp 124-125°C, $[\alpha]_D^{22}$ -39.6° (CH$_2$Cl$_2$, c 0.62) (Note 15).

    D. (2S)-(-)-3-(-)-exo-(Dimethylamino)isoborneol, [(-)-DAIB], 5. Caution! Since DAIB slowly decomposes in air, samples of DAIB should be stored under nitrogen or argon. All glassware for this step must be oven-dried. A 250-mL, three-necked, round-bottomed flask equipped with a reflux condenser connected to an argon inlet, a magnetic stirring bar, a rubber septum, and a Teflon stopper, is flushed with argon and charged with 55 mL of dry THF and 5.20 g (0.075 mol) of lithium aluminum hydride (Note 4). The solution is cooled to 0°C in an ice-water bath, and a solution of 2.77 g (0.013 mol) of N-methyl-(1R,2S,6R,7S)-1,10,10-trimethyl-4-oxo-5-aza-3-oxatricyclo[5.2.1.0]decane (**4**) in 35 mL of dry THF is introduced slowly with a cannula. After addition is complete, the flask is removed from the cooling bath, and the reaction mixture is heated at reflux for 18 hr. The reaction flask is cooled to 0°C in an ice-water bath, and the mixture is quenched by successive addition of 20 mL of ethyl acetate and 17 mL of saturated aqueous sodium sulfate solution (Notes 5 and 6). The white mixture is stirred at room temperature for 2 hr, and the resulting white granular precipitate or slurry is removed by vacuum filtration through Celite. The filter cake is washed with three 100-mL portions of ethyl acetate, and the combined filtrates are thoroughly dried over sodium sulfate, filtered, and concentrated by rotary evaporation to furnish crude **5** as an oil. Crude **5** is purified by bulb-to-bulb distillation to give 2.19 g (84%) of (2S)-(-)-3-exo-(dimethylamino)isoborneol, [(2S)-(-)-DAIB] (**5**) as a nearly colorless oil, bp 150°C (25 mm), $[\alpha]_D^{22}$ 36° (ethanol, c 4.65) (Notes 16 , 17, and 18).

2. Notes

1. Anhydrous, analytical reagent grade diethyl ether was purchased from Mallinckrodt Inc. and was used without additional drying.

2. Lithium aluminum hydride was purchased from Aldrich Chemical Company, Inc.

3. The oxime was prepared as described in *Org. Synth.* **2001**, *79*, 125.

4. Caution! A large volume of gas is produced.

5. This is an exothermic reaction and a large volume of gas is produced. Great care must be exercised at this stage to avoid adding the saturated aqueous sodium sulfate solution too rapidly.

6. In order to achieve efficient product recovery, it is important to add only sufficient saturated aqueous sodium sulfate solution to change the appearance of the reaction mixture from a gray slurry to a white slurry. At this point the white precipitate settles rapidly when stirring is stopped.

7. The spectral properties of (2S)-(-)-3-exo-aminoisoborneol (**2**) are as follows: ^{1}H NMR (300 MHz, CDCl$_3$) δ: 0.76 (s, 3 H), 0.79-1.03 (comp, 8 H), 1.35-1.42 (m, 1 H), 1.52 (d, 1 H, J = 4.6), 1.60-1.72 (m, 1 H), 2.4 (br s, 3 H), 3.02 (d, 1 H, J = 7.4), 3.34 (d, 1 H, J = 7.4); ^{13}C (125 MHz, CDCl$_3$) δ: 11.4, 21.2, 21.9, 26.9, 33.1, 46.6, 48.7, 53.4, 57.3, 79.0 ; IR (CHCl$_3$) cm^{-1}: 3279, 3019, 2957, 2884, 1478; mass spectrum (CI) m/z 170.1540 [C$_{10}$H$_{19}$NO (M+1) requires 170.1545], 170 (base), 166, 152.

8. Triphosgene (98%) was purchased from Aldrich Chemical Company, Inc.

9. Washing the organic layer with saturated sodium bicarbonate is accompanied by the vigorous evolution of large quantities of gas, so great care must be taken in this step to avoid loss of material.

10. This crude **3** can be purified by crystallization from a mixture of hexane and ethyl acetate to afford pure **3** as colorless crystals, mp 154-155°C (lit^2. mp 156-158°C), $[\alpha]_D^{22}$ -48° (CH$_2$Cl$_2$, c 0.63).

11. The spectral properties of **3** are as follows: ^{1}H NMR (300 MHz, CDCl$_3$) δ: 0.88 (s, 3 H), 1.01 (s, 3 H), 1.08 (s, 3 H), 1.45-1.74 (comp, 4 H), 1.84 (d, 1 H, J = 4.6), 3.75 (d, 1 H, J = 8.0), 4.37 (d, 1 H, J = 8.0 Hz), 5.83 (br s, 1 H); ^{13}C (75 MHz, CDCl$_3$) δ: 10.7, 19.3, 23.2, 24.7, 31.6, 46.3, 48.1, 48.6, 60.5, 88.3, 160.4; IR (CHCl3) cm^{-1}: 3470, 3268, 3019, 2962, 1746; mass spectrum (CI) m/z 196.1342 [C$_{11}$H$_{17}$NO$_2$ (M+1) requires 196.1338], 196 (base), 166, 89.

12. Potassium hydride (35 wt. % dispersion in mineral oil) was purchased from Aldrich Chemical Company, Inc.

13. Tetrahydrofuran (THF) was distilled from sodium-benzophenone ketyl under an atmosphere of argon immediately prior to use.

14. Methyl iodide was purchased from Aldrich Chemical Company, Inc.

15. The spectral properties of **4** are as follows: ^{1}H NMR (300 MHz, CDCl$_3$) δ: 0.87 (s, 3 H), 0.90-1.06 (comp, 8 H), 1.49-1.57 (m, 1 H), 1.70-1.79 (m, 1 H), 1.96 (d, 1 H, J = 4.6), 2.82 (s, 3 H), 3.55 (d, 1 H, J = 8.1), 4.22 (d, 1 H, J = 8.1); ^{13}C (75 MHz, CDCl$_3$) δ: 10.5, 18.8, 23.1, 24.6, 29.2, 31.7, 45.3, 46.1 48.2, 65.9, 84.6, 158.3; IR (CHCl$_3$) cm^{-1}: 3009, 2960, 1741, 1482, 1435, 1407; mass spectrum (CI) m/z 210.1491 [C$_{12}$H$_{19}$NO$_2$ (M+1) requires 210.1494], 210 (base).

16. The spectral properties of (2S)-(-)-exo-(dimethylamino)isoborneol, [(2S)-(-)-DAIB (**5**) are as follows: ^{1}H NMR (300 MHz, CDCl$_3$) δ: 0.74 (s, 3 H), 0.90-0.98 (m, 5 H), 1.02 (s, 3 H), 1.32-1.43 (m, 1 H), 1.62-1.71 (m, 1 H), 1.92 (d, 1 H, J = 4.6) 2.21 (d, 1 H, J = 7.0), 2.26 (br s, 6 H), 3.41 (d, 1 H, J = 7.0); ^{13}C (75 MHz, CDCl$_3$) δ: 11.5, 20.7, 22.0, 27.8, 32.2, 46.3, 47.0, 49.0, 74.1, 78.7 (only ten carbon signals are observed at 25°C due to line broadening of one of the N-methyl signals); IR (CHCl$_3$) cm^{-1}: 3310, 2991, 2956, 2875, 2830, 2784, 1469; mass spectrum (CI) m/z 198.1860 [C$_{12}$H$_{23}$NO (M+1) requires 198.1858], 198 (base).

17. This material is enantiomerically pure, as confirmed by HPLC analysis of the 3,5-dinitrophenyl carbamate using a chiral stationary phase. The carbamate derivative of DAIB

is prepared in the following manner: 3,5-Dinitrophenyl isocyanate (Sumika Chemical Analysis Service) and 5 mL of pyridine are added to a vigorously stirred solution of 10 mg (0.051 mmol) of DAIB (5) in 2 mL of toluene. After stirring for 30 min at 20°C, the reaction mixture is passed through 0.5 g of Merck 9385 silica gel, eluting with diethyl ether. The eluent is concentrated by rotary evaporation to give the crude N-3,5-dinitrophenyl carbamate. This material is analyzed by HPLC (Shimadzu LC-6A chromatograph equipped with a Rheodyne 7125 injector and a Shimadzu SPD-6A UV detector. A Sumitomo Chemical Co. Sumipax OA-4000 column is employed and the carbonate is eluted with a 99.5:0.5 mixture of hexane and isopropyl alcohol (flow rate, 1.0 mL/min; detection, 254 nm; tR, 23.7 min). The signal due to the minor R isomer could not be detected under the analysis conditions, and the ee of the sample is estimated to be at least 99.9%. (We are indebted to Professor Ryoji Noyori for this analysis.)

18. Toward the end of the distillation, a small amount of colored material begins to distill over, so the distillation must be stopped at this point.

Waste Disposal Information

All toxic materials were disposed of in accordance with "Prudent Practices in the Laboratory"; National Academy Press; Washington, DC, 1995.

3. Discussion

The chiral β-dialkylamino alcohol DAIB serves as an efficient asymmetric catalyst for the addition of organozinc reagents to aldehydes.[3] The reaction of diethylzinc with benzaldehyde in the presence of 2 mol % of (2S)-(-)-DAIB to give (S)-1-phenyl-1-propanol in 89% ee is described in the procedure which follows.[4] DAIB exhibits the property of "chiral amplification," enabling the preparation of alcohols of high enantiomeric purity to be effected with catalyst of much lower enantiomeric purity.

Conversion of the keto ketoxime **1** to the exo-exo-amino alcohol **2** has been accomplished by hydrogenation over Adams' catalyst[5] and by reduction with lithium aluminum hydride.[6] Amino alcohol **2** has also been prepared from **1** by a two-stage process in which selective reduction of the ketone is carried out with sodium borohydride, and the resultant hydroxy oxime is reduced with lithium aluminum hydride[6] or by hydrogenation over Adams catalyst.[7]

Conversion of **2** to the highly crystalline oxazolidinone **3** with phosgene has been described by Thornton[7] who has employed this substance as a chiral auxiliary in asymmetric aldol reactions of its N-propionyl derivative. Kelly has also used an oxazoline derived from **3** as a chiral auxiliary in asymmetric alkylation of a glycolate enolate.[8] Oxazolidinone **3** has also been prepared from **2** with diethyl carbonate in the presence of potassium carbonate.[2] The conversion of **2** to the oxazolidinone **3** is accomplished using triphosgene in this procedure because of the high toxicity of phosgene.

N-Methylation of **3** and reduction of the crystalline oxazolidinone **4** with lithium aluminum hydride was found to give a superior yield of DAIB (**5**) and a more easily purified product than exhaustive methylation of **2** with methyl iodide and reduction of the quaternary methiodide with Super-Hydride.[9] Recently, a modified version of DAIB, 3-exo-morpholinoisoborneol MIB), was prepared by Nugent that is crystalline and that is reported to give alcohols in high enantiomeric excess from the reaction of diethylzinc with aldehydes.[10]

1. Department of Chemistry, Oregon State University, Corvallis, OR 97331-4003.

2. Tanaka, K.; Ushio, H.; Kawabata, Y.; Suzuki, H. *J. Chem. Soc., Perkin Trans. I* **1991**, 1445.

3.   Noyori, R. "Asymmetric Catalysis in Organic Syntheses", Wiley-Interscience: New York, 1994; pp. 261-278.

4.   Kitamura, M.; Oka, H.; Suga, S.; Noyori, R. *Org. Synth.* **2001**, *79*, 139.

5.   Beckett, A. H.; Lan, N. T.; McDonough, G. R. *Tetrahedron* **1969**, *25*, 5689.

6.   Chittenden, R. A.; Copper, G. H. J. *Chem. Soc. (C)* **1970**, 49.

7.   Bonner, M. P.; Thornton, E. R. *J. Am. Chem. Soc.* **1991**, *113*, 1299.

8.   Kelly, T. R.; Arvanitis, A. *Tetrahedron Lett.* **1984**, *25*, 39.

9.   Oppolzer, W.; DeBrabander, J., private communication.

10. Nugent, W. A. *J. Chem. Soc.,Chem. Commun.* **1999**, 1369

Appendix

Chemical Abstracts Nomenclature (Collective Index Number); (Registry Number)

(2S)-(-)-3-exo-(Dimethylamino)isoborneol: (2S)-(-)-DAIB: Bicyclo[2.2.1]heptan-2-ol, 3-(dimethylamino)-1,7,7-trimethyl-, [1R-(exo,exo)]- (11); (103729-96-0)

(2S)-(-)-3-exo-Aminoisoborneol: Bicyclo[2.2.1]heptan-2-ol, 3-amino-1,7,7-trimethyl-, (1R,2S,3R,4S)- (9); (41719-73-7)

Lithium aluminum hydride: Aluminate (1-), tetrahydro-, lithium (8); Aluminate (1-), tetrahydro-, lithium, (T-4)- (9); (16853-85-3)

(1R,4S0-(-)-Camphorquinone monoxime: 2,3-Bornanedione, 3-oxime (9); Bicyclo[2.2.1]heptane-2,3-dione, 1,7,7-trimethyl-, 3-oxime (12); (663-17-2)

(1R,2S,6R,7S)-1,10,10-Trimethyl-4-oxo-5-aza-3-oxatricyclo[5.2.1.0]decane: 4,7-Methanobenzoxazol-2(3H)-one, hexahydro-7,8,8-trimethyl-, [3aR-(3aα,4β,7β,7aα)]- (12); (131897-87-5)

Triphosgene: Carbonic acid, bis(trichloromethyl) ester (8,9); (32315-10-9)

Potassium hydride (8,9); (7693-26-7)

Methyl iodide: Methane, iodo- (8,9); (74-88-4)

CATALYTIC ENANTIOSELECTIVE ADDITION OF DIALKYLZINCS TO ALDEHYDES USING (2S)-(-)-3-exo-(DIMETHYLAMINO)ISOBORNEOL [(2S)-DAIB]: (S)-1-PHENYL-1-PROPANOL

(Benzenemethanol, α-ethyl-, (S)-)

Submitted by Masato Kitamura, Hiromasa Oka, Seiji Suga, and Ryoji Noyori.[1]
Checked by David E. Kaelin, Stephen F. Martin, Gregory L. Beutner and Scott E. Denmark.

1. Procedure

Caution! Since DAIB deteriorates in air, bottles of DAIB should be flushed with N_2 or Ar and kept tightly closed for storage over long periods. Diethylzinc easily catches fire upon contact with air or moisture and addition to benzaldehyde should be performed under anaerobic conditions using degassed solvent.

A dry, 500-mL Schlenk tube, equipped with a rubber septum and a Teflon-coated stirring bar and filled with argon (Note 1) is charged with (2S)-(-)-3-exo-(dimethylamino)isoborneol [(2S)-DAIB] (371 mg, 1.88 mmol) (Note 2), dry toluene (200 mL) (Note 3), and a 4.45 M toluene solution of diethylzinc (25.4 mL, 113 mmol) through a rubber septum using hypodermic syringes at 20°C (Notes 4 and 5). The mixture is stirred for 15 min and then cooled to -78°C with a dry ice-methanol bath. To this is added benzaldehyde (10.0 g, 94.2 mol) (Note 6) in one portion (Note 7). The bath is replaced by an ice bath, and the septum is replaced by a glass stopper. The reaction mixture is stirred at

0°C for 6 hr in a closed system. The glass stopper is removed under an argon stream, and saturated aqueous ammonium chloride solution (40 mL) is carefully added (Note 5), resulting in the formation of a white precipitate. The liquid layer and the solid phase are roughly separated by decantation. The precipitate is washed with ether (100 mL), and the combined liquid layers and 2 M aqueous hydrochloric acid solution (100 mL) are transferred to a 1-L separatory funnel. The aqueous layer is separated and extracted twice with ether (100 mL) (Note 8). The combined organic layers (ca. 550 mL) are washed with water (50 mL) and brine (50 mL), dried over anhydrous sodium sulfate, and concentrated under reduced pressure. The crude residue is distilled at 150-155°C and 20 mm Hg using a Kugelrohr apparatus to give 12.4 g of (S)-1-phenyl-1-propanol (97% yield) in 95.4% ee as a colorless oil (Notes 9 and 10).

2. Notes

1. Argon gas (99.998%) is used without further purification. The Schlenk tube and syringes are dried overnight at 150°C.

2. Supplied by Professor James D.White Oregon State University[2]; see previous procedure, p.130.

3. Toluene is first distilled from sodium benzophenone ketyl under argon. Prior to the reaction, the toluene is degassed by performing two freeze-pump-thaw cycles, then the flask is backfilled with argon.

4. A stock solution of diethylzinc is prepared in an 80-mL Schlenk tube equipped with a Young's tap by mixing toluene and 99% diethylzinc [(16.5 g, 134 mmol) in a lecture bottle, which is purchased from Aldrich Chemical Company, Inc.] to make a total volume of 30 mL.

5. Ethane gas evolution is observed.

6. Benzaldehyde, which is purchased from Aldrich Chemical Company, Inc., is purified by distillation from 4 Å molecular sieves (70.5-71.5°C/20 mmHg) and kept in a Schlenk tube equipped with a Young's tap.

7. The solution becomes a pale yellow color; the color fades after 6 hr.

8. (2S)-DAIB (312 mg) is recovered in 77–84% yield from the aqueous layer. To the aqueous layer (pH <2), cooled with an ice bath, is added 6 M aqueous sodium hydroxide solution (60 mL). The mixture, which has a pH of >12, is extracted three times with ether (50 mL). The combined organic layers are washed with water (20 mL) and brine (20 mL), dried over anhydrous sodium sulfate, and concentrated under reduced pressure to give a colorless oil.

9. The product, which contains 2-3% of benzyl alcohol and 2-3% of propiophenone, has the following physical properties: $[\alpha]_D^{21}$ −45.6° CHCl$_3$, c 5.55); ^{1}H NMR (400 MHz, CDCl$_3$) δ: 0.91 (t, 3, J = 7.3, CH$_3$), 1.68-1.88 (m, 2, CH$_2$), 1.90-1.95 (m, 1, OH), 4.58 (td, 1, J = 6.4, 3.4, OCH), 7.21-7.38 (5, aromatic protons); ^{13}C NMR (100 MHz, CDCl$_3$) δ: 10.1, 31.9, 76.0, 125.9, 127.5, 128.4, 144.6.

10. The enantiomeric purity is determined by chiral stationary phase, supercritical fluid chromatographic (CSP-SFC) analysis (Berger Instruments, Daicel Co. CHIRALCEL OD column; 4% methanol, 180 psi, 3.0 mL/min flow rate; detection at 220 nm). Racemic 1-phenylpropanol exhibited base-line separation of peaks of equal intensity arising from the R-isomer (t_R 2.74 min) and the S-isomer (t_R 3.10 min) whereas the synthetic alcohol showed these peaks in the ratio 97.7 / 2.3. This chromatographic method allowed for identification of the trace contaminants propiophenone (t_R 1.63 min) and benzyl alcohol (t_R 3.40 min).

11. The submitters used HPLC analysis to determine the enantiomeric purity (Daicel Co. CHIRALCEL OB column; 99.5/0.5 hexane/2-propanol mixture, 1.0 mL/min flow rate; detection at 254-nmS-isomer t_R 19.2 min, R-isomer t_R 24.6 min)). Under these conditions, however, the propiophenone contaminant is coincident with the S-enantiomer thus affording unreliable enantiomeric analysis.

141

Waste Disposal Information

All toxic materials were disposed of in accordance with "Prudent Practices in the Laboratory"; National Academy Press; Washington, DC, 1995.

3. Discussion

(2S)-DAIB, a chiral β-dialkylamino alcohol, serves as an efficient catalyst for enantioselective addition of diethylzinc to benzaldehyde in toluene, hexane, ether, or their mixtures, giving (S)-1-phenyl-1-propanol in up to 99% ee.[3,4] This catalytic enantioselective alkylation can be extended to a range of alkylating agents and aldehyde substrates, as illustrated in the Table. Dimethyl-, diethyl-, and other simple dialkylzinc agents can be used as alkylating agents. p-Substituted benzaldehydes give the corresponding secondary alcohols with consistently high enantioselectivity. 2-Furaldehyde is alkylated with di-n-pentylzinc in the presence of (2S)-DAIB to give (1S)-1-(2-furyl)hexan-1-ol, a versatile compound in organic synthesis, with >95% ee. Optically active 1-ferrocenylethanol, a key compound for the synthesis of a wide variety of chiral ferrocene derivatives, is obtained in 81% optical yield by methylation of ferrocenecarboxaldehyde. Certain α,β-unsaturated and aliphatic aldehydes can also be alkylated in moderate to high optical yield. The (2S)-DAIB-catalyzed addition of di-n-pentylzinc to (E)-3-tributylstannylpropenal proceeded with an S:R selectivity of 93:7 to afford a key chiral building block that was used in the three-component coupling step of a prostaglandin synthesis. The DAIB-catalyzed reaction of (1-alkenyl)ethylzincs, prepared by transmetalation of (1-alkenyl)dicyclohexylboranes with diethylzinc, plays a key role in the asymmetric syntheses of muscone and aspicilin.[5] Polystyrene-anchored DAIB can also be used as a chiral auxiliary for enantioselective reactions.[6]

(2S)-DAIB-PROMOTED ENANTIOSELECTIVE ADDITION OF DIORGANOZINCS TO ALDEHYDES[3,4]

Aldehyde	Diorganozinc	Product	% Yield	% ee
X = H	$(CH_3)_2Zn$	R = CH_3	95	95
	$(C_2H_5)_2Zn$	R = C_2H_5	97	98
	$(n\text{-}C_4H_9)_2Zn$	R = $n\text{-}C_4H_9$	88	95
X = Cl	$(C_2H_5)_2Zn$	R = C_2H_5	86	93
X = CH_3O	$(C_2H_5)_2Zn$	R = C_2H_5	96	93
	$(n\text{-}C_5H_{11})_2Zn$		80	>95
	$(CH_3)_2Zn$		60	81
	$(C_2H_5)_2Zn$		81	96
	$(n\text{-}C_5H_{11})_2Zn$		84	85
	$(C_2H_5)_2Zn$		80	90

143

75% yield
92% ee

Dramatic chiral amplification is observed in alkylations catalyzed by partially resolved DAIB.[7] Reaction of benzaldehyde and diethylzinc in toluene containing 8 mol % of (2S)-DAIB of 15% ee leads to (S)-1-phenyl-1-propanol in 95% ee, a value close to the 98% ee obtained with enantiomerically pure (2S)-DAIB.

A combined system consisting of a Ni(II) complex and (2S)-DAIB effects the conjugate addition of diethylzinc to chalcone, resulting in the formation of (R)-1,3-diphenylpenta-1-one in 85% ee.[8]

+ $(C_2H_5)Zn$

7% Ni(acac)$_2$/2,2'-bipyridine
17% (2S)-DAIB
CH_3CN

75% yield
85% ee

1. Department of Chemistry and Molecular Chirality Research Unit, Nagoya University, Chikusa, Nagoya 464-01, Japan.

2. White, J.; Wardrop, D. J.; Sundermann, K. F. *Org. Synth.* **2001**, *79*, 130.

3. Kitamura, M.; Suga, S.; Kawai, K.; Noyori, R. *J. Am. Chem. Soc.* **1986**, *108*, 6071; Noyori, R.; Suga, S.; Kawai, K.; Okada, S.; Kitamura, M.; Oguni, N.; Hayashi, M.; Kaneko, T.; Matsuda, Y. *J. Organomet. Chem.* **1990**, *382*, 19.

4. Noyori, R.; Kitamura, M. *Angew. Chem., Int. Ed. Engl.* **1991**, *30*, 49.

5. Oppolzer, W.; Radinov, R. N. *J. Am. Chem. Soc.* **1993**, *115*, 1593; Oppolzer, W.; Radinov, R. N.; De Brabander, J. *Tetrahedron Lett.* **1995**, *36*, 2607.

6. Itsuno, S.; Fréchet, J. M. J. *J. Org. Chem.* **1987**, *52*, 4140.

7. Noyori, R.; Suga, S.; Kawai, K.; Okada, S.; Kitamura, M. *Pure Appl. Chem.* **1988**, *60*, 1597; Kitamura, M.; Okada, S.; Suga, S.; Noyori, R. *J. Am. Chem. Soc.* **1989**, *111*, 4028; Kitamura, M.; Suga, S.; Niwa, M.; Noyori, R.; Zhai, Z.-X.; Suga, H. *J. Phys. Chem.* **1994**, *98*, 12776; Kitamura, M.; Suga, S.; Niwa, M.; Noyori, R. *J. Am. Chem. Soc.* **1995**, *117*, 4832; Yamakawa, M.; Noyori, R. *J. Am. Chem. Soc.* **1995**, *117*, 6327; Kitamura, M.; Yamakawa, M.; Oka, H.; Suga, S.; Noyori, R. *Chem. Eur. J.* **1996**, *2*, 1173. For the original discovery of this effect, see: Oguni, N.; Matsuda, Y.; Kaneko, T. *J. Am. Chem. Soc.* **1988**, *110*, 7877.

8. Jansen, J. F. G. A.; Feringa, B. L. *Tetrahedron: Asymmetry* **1992**, *3*, 581; De Vries, A. H. M.; Jansen, J. F. G. A.; Feringa, B. L. *Tetrahedron* **1994**, *50*, 4479.

Appendix
Chemical Abstracts Nomenclature (Collective Index Number);
(Registry Number)

(2S)-3-exo-Aminoisoborneol: Bicyclo[2.2.1]heptan-2-ol, 3-amino-1,7,7-trimethyl-, [1R-(exo,exo)]; (41719-73-7)

(2S)-3-exo-(Dimethylamino)isoborneol: Bicyclo[2.2.1]heptan-2-ol, 3-(dimethylamino)-1,7,7-trimethyl-, [1R-(exo,exo)]; (103729-96-0)

(S)-1-Phenyl-1-propanol: Benzenemethanol, .alpha.-ethyl-, (S)-; (613-87-6)

(R)-1-Phenyl-1-propanol: Benzenemethanol, .alpha.-ethyl-, (R)-; (1565-74-8)

(±)-1-Phenyl-1-propanol: Benzenemethanol, .alpha.-ethyl-, (93-54-9)

Benzaldehyde: Benzaldehyde; (100-52-7)

FORMATION OF γ-KETO ESTERS FROM β-KETO ESTERS:

METHYL 5,5-DIMETHYL-4-OXOHEXANOATE

(Hexanoic acid, 5,5-dimethyl-4-oxo-, methyl ester)

Submitted by Matthew D. Ronsheim, Ramona K. Hilgenkamp, and Charles K. Zercher.[1]

Checked by Scott E. Denmark and Gregory L. Beutner.

1. Procedure

CAUTION! Neat diethylzinc may ignite on exposure to air and reacts violently with water. It must be handled and reacted under nitrogen.[2] The reaction solvent must be dried and distilled prior to use and all glassware and syringes must be thoroughly dried.

An oven-dried, 3-L, three-necked, round-bottomed flask equipped with a magnetic stirring bar is charged with 1000 mL of freshly distilled methylene chloride (Note 1). One of the outer two necks of the flask is equipped with a gas inlet adapter attached to a nitrogen source and the other with a septum connected through a needle to a bubbler (Notes 2,3). A 125-mL pressure-equalizing addition funnel capped with a septum is placed in the center neck of the flask with the stopcock closed. The funnel is charged with 12.1 mL (150 mmol) of methylene iodide (Note 4) dissolved in 75 mL of distilled methylene chloride. The solution is stirred under nitrogen for 1 hr and the flask is cooled in an ice-water bath. Using a syringe, 15.4 mL (150 mmol) of neat diethylzinc (Note 5) is added to the methylene chloride through the septum (Note 6). The stopcock of the addition funnel is opened carefully and the methylene iodide solution is allowed to drip into the diethylzinc solution over the course of 30 min (Note 7). After the reaction is stirred for an additional 10 min, 8.0 mL (50 mmol) of methyl 4,4-dimethyl-3-oxopentanoate is added to the flask over a period of 15 sec through the septum using a syringe (Note 8). The reaction mixture is stirred for 45 min (Note 9), the septum is removed, and the reaction is quenched through the careful addition of 125 mL of saturated aqueous ammonium chloride (Note 10,11). Stirring is maintained for an additional 10 min, at which time the reaction mixture is transferred to a 2-L separatory funnel. The lower organic layer is withdrawn and placed in a large, round-bottomed flask. The methylene chloride is removed using a rotary evaporator (Note 12) and the residue is diluted with 500 mL of diethyl ether and placed in the separatory funnel. It is washed successively with 125-mL portions of saturated aqueous sodium bicarbonate, 1 M aqueous sodium thiosulfate solution (Note 13), deionized water and brine. The aqueous washings are extracted with diethyl ether (3 x 150 mL) and the combined organic layers are dried over 50 g of anhydrous sodium sulfate prior to concentration under reduced pressure. The product is purified by vacuum distillation from anhydrous potassium carbonate (58°C, 1 mm) to give methyl 5,5-dimethyl-4-oxo-hexanoate (7.63-8.09 g, 89-94% yield) as a clear liquid (Notes 14, 15).

2. Notes

1. Methylene chloride, reagent grade, was obtained from Pharmco Products, Inc. and was distilled from phosphorus pentoxide prior to use. Benzene can be substituted for methylene chloride in the chain extension reaction with similar results; however, the use of benzene in the reaction has not been independently checked.

2. The first step of the chain extension reaction mechanism has been shown to be enolate formation, the by-product of which is ethane gas. The nitrogen line should be attached to a large adapter to provide adequate venting of the gas.

3. The checkers used a sidearm vacuum adapter fitted with a septum and connected to a nitrogen manifold. The third neck was stoppered.

4. Methylene iodide was obtained from Lancaster Synthesis, Inc., and used without further purification. Distillation of the methylene iodide from copper prior to use does not lead to increased product yield or purity.

5. Three equivalents of the carbenoid ($EtZnCH_2I$) are necessary for the clean formation of the product. One of the equivalents serves to facilitate formation of the enolate, while a second equivalent provides the methylene group that is incorporated into the product. However, if the reaction is performed with only two equivalents, significant amounts of starting material and a second product, methyl 2,4,4-trimethyl-3-oxopentanoate, are generated in the reaction.

6. Diethylzinc was obtained from Aldrich Chemical Company, Inc., and used without further purification. If the reaction flask has been carefully flushed with nitrogen, there should be very little smoke evident when the zinc reagent is added. Diethylzinc can be added with the syringe tip placed below the surface of the solvent rather than above it.

7. Formation of the zinc carbenoid is exothermic and potentially explosive.[3] The ice bath should be present in order to control the reaction temperature, and methylene iodide should be added gradually rather than all at once. During the addition of the methylene iodide, or shortly thereafter, the reaction mixture should develop a cloudy, white appearance.

8. Methyl 4,4-dimethyl-3-oxopentanoate was obtained from Lancaster Synthesis, Inc., and used without further purification.

9. Disappearance of the starting material may be monitored by thin layer chromatography (R_f = 0.6, 2:1 hexane:ethyl acetate). It is important that the reaction time not be extended past the point when all of the β-keto ester is consumed, because long exposure to the zinc carbenoid has been shown to result in substrate decomposition and reduced yields.

10.   Addition of the aqueous solution is exothermic and should, therefore, be carried out over the course of a few minutes. The checkers found that the initial internal temperature was 5-6°C which rose to 10°C upon addition of the first few mL of quench solution.

11.   The checkers found that addition of 150 mL of 2 N aqueous hydrochloric acid rather than 125 mL of saturated aqueous ammonium chloride does not alter the yield or purity of the product.

12.   Methylene chloride is exchanged for diethyl ether, since removal of the zinc salts formed as reaction by-products is facilitated by use of an extraction solvent that is less dense than water.

13.   The checkers found that washing the organic extracts with sodium thiosulfate solution was necessary to obtain a colorless product.

14.   The NMR spectra were as follows and were consistent with those reported in the literature:[4] ^{1}H NMR (500 MHz, CDCl$_3$) δ: 1.16 (s, 9 H), 2.57 (t, 2 H, J = 6.36), 2.81 (t, 2 H, J = 6.36), 3.67 (s, 3 H); ^{13}C NMR (125 MHz, CDCl$_3$) δ: 26.4, 27.8, 31.4, 43.9, 51.7, 173.5, 214.4; IR (neat) 2970, 2874, 1741, 1707, 1366, 1204, 1167, 1087. MS: (EI) 172.1 (0.23), 155.1 (0.43), 141.1 (16), 115.0 (100), 57.1 (64). Anal. Calcd. for C$_9$H$_{16}$O$_3$: C, 62.75; H, 9.36. Found: C, 62.42; H, 9.38

15.   The checkers found that distillation from anhydrous potassium carbonate is required in order to obtain a clear, analytically pure sample. Simple distillation yields a slightly yellow liquid which is free from major contamination, but not analytically pure.

Waste Disposal Information

All toxic materials were disposed of in accordance with "Prudent Practices in the Laboratory"; National Academy Press; Washington, DC, 1995.

3. Discussion

A variety of methods have been reported for the preparation of γ-keto esters. Although strategies in which the ketone and ester functionalities are assembled from different sources are frequently used,[5] reactions that promote the insertion of a single methylene unit between the carbonyl functionalities of readily accessible β-keto esters are very attractive. The formation and fragmentation of 2-carboxycyclopropyl alcohols have been central to this methodological development.[6]

Three complementary strategies have been implemented for the formation of γ-keto esters through these functionalized cyclopropyl alcohols.[7] While each of the three methods provided access to γ-keto esters, the multiple steps and/or poor yields of the transformations serve to discourage their widespread applicability. The submitters have developed a simple and efficient one-step method for the formation of γ-keto esters from β-keto esters that appears to proceed through a similar cyclopropyl alcohol intermediate.[8] Exposure of an α-unsubstituted β-keto ester to a 1:1 mixture of diethylzinc and methylene iodide results in its clean and rapid conversion to the chain-extended keto ester. This zinc-mediated process has two distinct advantages over the previously reported chain-extension methods. The most obvious advantage is that no additional steps are required for the formation of the intermediate enol ether or for the cleavage of the protected cyclopropyl alcohol. Secondly, utilization of diethylzinc is operationally much simpler than preparation and application of the zinc-copper couple reported by Saigo.[7h]

The zinc-mediated reaction tolerates a variety of functionality in the β-keto ester. In fact, the method described above has been applied successfully to β-keto amides[9] and β-keto phosphonates.[10] Unsubstituted β-keto esters, amides and phosphonates have been chain-extended in yields that ranged from 58% to 98% (Table I). The primary limitation to this method is the inefficiency with which α-substituted esters and amides undergo methylene insertion. The zinc carbenoid must be employed in at least a threefold excess in order to fully convert β-keto ester to products and to avoid the formation of α-methylated β-keto esters. Side products that result from an intermolecular reaction are observed on occasion. For example, the preparation of ethyl 4-oxo-4-phenylbutanoate from ethyl benzoylacetate on a 50-mmol scale proceeded in only 70% isolated yield because of a competing intermolecular reaction.

TABLE
PRODUCTS OF ZINC CARBENOID-MEDIATED CHAIN EXTENSION REACTIONS

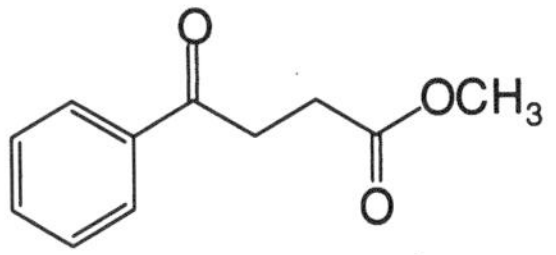

70%[a] (58%)[b,c]

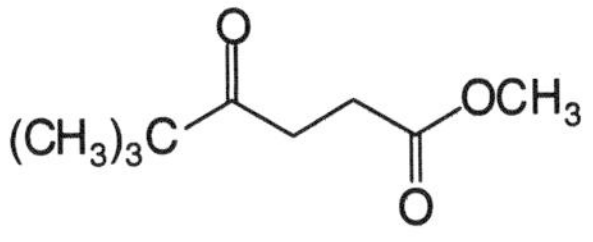

89-94% (71%)[b,c]

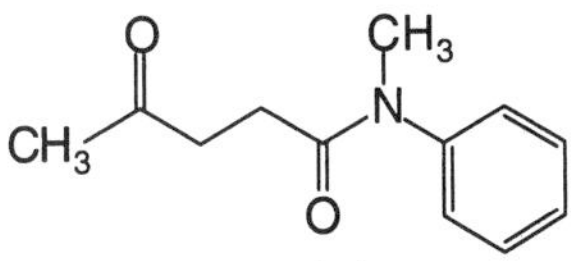

88%[b,d]

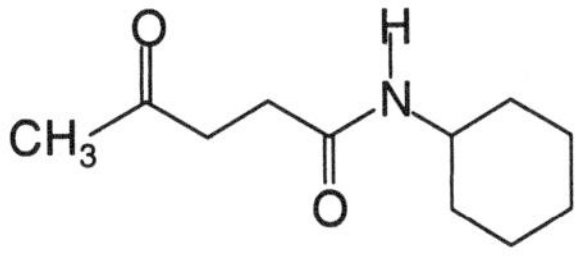

81%[b,d]

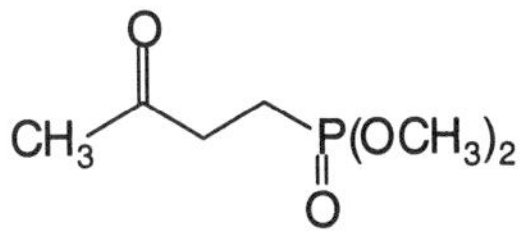

85%[b,e]

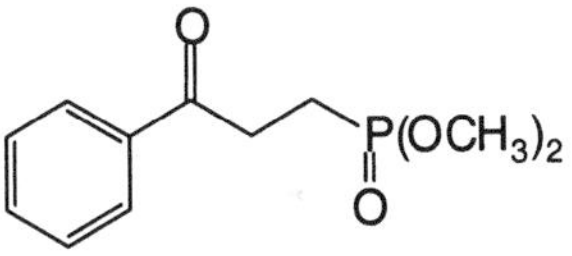

98%[b,e]

(a) 50-mmol scale with 4 equiv of carbenoid; (b) isolated yield after chromatographic purification of a 1-mmol scale reaction; (c) 5 equiv of carbenoid; (d) 4 equiv of carbenoid; (e) 6 equiv of carbenoid.

1. Department of Chemistry, University of New Hampshire, Durham, NH 03824.

2. Patnaik, P., "Comprehensive Guide to the Hazardous Properties of Chemical Substances," Van Nostrand Reinhold: New York, 1992.

3. (a) Furukawa, J.; Kawabata, N.; Nishimura, J. *Tetrahedron* **1968**, *24*, 53; (b) Charette, A. B.; Prescott, S.; Brochu, C. *J. Org. Chem.* **1995**, *60*, 1081.

4. (a) Marks, M. J.; Walborsky, H. M. *J. Org. Chem.* **1981**, *46*, 5405; (b) Hill G. A.; Salvin, V.; O'Brien, W. T. M. *J. Am. Chem. Soc.* **1937**, *59*, 2385.

5. (a) Miyakoshi, T. *Org. Prep. Proced. Int.* **1989**, *21*, 659; (b) Nagata, W.; Yoshioka, M. *Org. React.* **1977**, *25*, 255.

6. Reissig, H. -U. *Top. Curr. Chem.* **1988**, *144*, 73.

7. (a) Wenkert, E.; McPherson, C. A.; Sanchez, E. L.; Webb, R. L. *Synth. Commun.* **1973**, *3*, 255; (b) Reichelt, I.; Reissig, H. -U. *Chem. Ber.* **1983**, *116*, 3895; (c) Kunkel, E.; Reichelt, I.; Reissig, H. -U. *Liebigs Ann. Chem.* **1984**, 802; (d) Reichelt, I.; Reissig, H. -U. *Leibigs Ann. Chem.* **1984**, 531; (e) Saigo, K.; Kurihara, H.; Miura, H.; Hongu, A.; Kubota, N.; Nohira, H.; Haseqawa, M. *Synth. Commun.* **1984**, *14*, 787; (f) Grimm, E. L.; Reissig, H. -U. *J. Org. Chem.* **1985**, *50*, 242; (g) Bieräugel, H.; Akkerman, J. M.; Lapierre Armande, J. C.; Pandit, U. K. *Tetrahedron Lett.* **1974**, 2817; (h) Saigo, K.; Yamashita, T.; Hongu, A.; Hasegawa, M. *Synth. Commun.* **1985**, *15*, 715; (i) Dowd, P.; Choi, S. -C. *J. Am. Chem. Soc.* **1987**, *109*, 3493; (j) Dowd, P.; Choi, S. -C. *J. Am. Chem. Soc.* **1987**, *109*, 6548; (k) Dowd, P.; Choi, S. -C. *Tetrahedron* **1989**, *45*, 77; (l) Dowd, P.; Choi, S. -C. *Tetrahedron Lett.* **1989**, *30*, 6129.

8. Brogan, J. B.; Zercher, C. K. *J. Org. Chem.* **1997**, *62*, 6444.

9. Hilgenkamp, R.; Zercher, C. K. *Tetrahedron* **2001**, *57*, 8793

10. Verbicky, C. A.; Zercher, C. K. *J. Org. Chem.* **2000**, *65*, 5615.

Appendix
Chemical Abstracts Nomenclature (Collective Index Number);
(Registry Number)

Methyl 5,5-dimethyl-4-oxohexanoate: Hexanoic acid, 5,5-dimethyl-4-oxo-, methyl ester (9); (34553-32-7)

Diethylzinc: FLAMMABLE LIQUID: Zinc, diethyl- (8, 9); (557-20-0)

Methylene iodide: Methane, diiodo- (8, 9); (75-11-6)

Methyl 4,4-dimethyl-3-oxopentanoate: Pentanoic acid, 4,4-dimethyl-3-oxo-, methyl ester (9); (55107-14-7)

(S)-3-(tert-BUTYLOXYCARBONYLAMINO)-4-PHENYLBUTANOIC ACID

[Benzenebutanoic acid, β-[[(1,1-dimethylethoxy)carbonyl]amino]-, (S)-

Submitted by Michael R. Linder, Steffen Steurer, and Joachim Podlech.[1]

Checked by Frédéric Berst and Andrew B. Holmes.

1. Procedure

Caution! Diazomethane should be handled in an efficient fume hood behind a protection shield because of its toxicity and the possibility of explosions.

A. *(S)-3-(tert-Butyloxycarbonylamino)-1-diazo-4-phenylbutan-2-one.* A 1-L, three-necked, round-bottomed flask is equipped with a magnetic stirring bar, nitrogen gas inlet, bubble counter and a rubber septum on the center neck. The apparatus is dried under a rapid stream of nitrogen with a heat gun. After the flask is cooled to room temperature, the rate of nitrogen flow is reduced and Boc-phenylalanine (25.0 g, 94.2 mmol, Note 1) and anhydrous tetrahydrofuran (250 mL, Note 2) are added. The flask is immersed in an ice-water bath and triethylamine (13.1 mL, 94.0 mmol, Note 3) is added. After 15 min ethyl chloroformate (9.45 mL, 94.0 mmol, Note 4) is added. The reaction mixture is stirred for another 15 min, and a white precipitate of triethylammonium chloride appears; the stirring is then stopped. The septum is replaced by a funnel (Note 5). An ethereal solution of diazomethane (about 125 mL, Note 6) is added through the funnel, stirring is resumed for about 5 seconds and the nitrogen stream is stopped. After 45 min, the remainder of the diazomethane solution (about 85 mL) is added. The cooling bath is removed and the solution is allowed to react for 3 hr without stirring. With stirring, 75 mL of 0.5 N acetic acid is added carefully to destroy unreacted diazomethane and saturated aqueous sodium bicarbonate solution (75 mL) is added carefully. The aqueous layer is separated in a separatory funnel and the organic layer is washed with saturated aqueous sodium chloride (75 mL). The organic layer is dried over magnesium sulfate, filtered, and the solvents are removed under vacuum on a rotary evaporator. The crude product is placed under high vacuum for 3 hr (Note 7). The crude material is used directly in the next step (Notes 8,9).

B. *(S)-3-(tert-Butyloxycarbonylamino)-4-phenylbutanoic acid.* A 500-mL, three-necked flask is equipped with a nitrogen gas inlet, bubble counter, septum and a magnetic stirring bar. The flask is carefully wrapped in aluminum foil (to exclude light during the reaction). The crude diazo ketone from the preceding step is dissolved in tetrahydrofuran (380 mL, Note 10) and added to the flask under an atmosphere of nitrogen. De-ionized water (38 mL) is added, the flask is immersed in a dry ice-acetone bath, and the solution is cooled to – 25°C (temperature of the acetone cooling bath) for 30 min. Silver trifluoroacetate (2.72 g, 12.3 mmol, Note 11) is placed in a 50-mL Erlenmeyer flask and quickly dissolved in triethylamine (39 mL, 279 mmol, Note 3). The resulting solution is added to the diazo ketone solution in one portion (via syringe). The solution is allowed to

155

warm to room temperature overnight. Evolution of nitrogen starts at a bath temperature of about −15°C.

The solution is transferred to a 1-L, round-bottomed flask and the reaction vessel is rinsed with ethyl acetate (2 x 10 mL). The solution is evaporated to dryness with a rotary evaporator and the residue is stirred for 1 hr with saturated aqueous sodium bicarbonate (NaHCO$_3$) solution (100 mL, Note 12). The black mixture is transferred into a 1-L separatory funnel with water (150 mL) and ethyl acetate (200 mL), and the mixture is shaken well. The clear aqueous layer is separated and put aside, leaving an organic phase containing a suspension of black solid. Brine (30 mL) is added to the organic phase and the resulting mixture is shaken vigorously. Saturated, aqueous NaHCO$_3$ solution (30 mL) is added, the medium is shaken again, and the layers are separated. The black solid is carried away with the aqueous phase, which is now combined with the first-separated aqueous phase. The organic layer is washed with three additional portions of saturated aqueous NaHCO$_3$ solution (30 mL each) and all the aqueous layers are combined. The first organic layer is put aside and not used further. The combined aqueous layers containing a black suspension are extracted with ethyl acetate (50 mL) and the ethyl acetate layer is then back-extracted with two portions of saturated aqueous NaHCO$_3$ solution (25 mL each), which are combined with the original aqueous layers. The ethyl acetate is put aside and not used further. All the combined aqueous layers are extracted again with 50 mL of ethyl acetate, which is washed with saturated aqueous NaHCO$_3$ solution (2 x 20 mL, Note 13). The organic layer is put aside and not used further. All the combined aqueous layers are then transferred to a 2-L, round-bottomed flask equipped with a magnetic stirring bar and about 10 drops of Congo Red indicator (Note 14) and ethyl acetate (100 mL) are added. The flask is immersed in an ice-water bath, the solution is stirred and 5 N (17.5 wt %) hydrochloric acid is added dropwise through an addition funnel until the color of the indicator changes from red to blue (Note 15). The solution is placed in a 1-L separatory funnel and the organic layer is separated. The aqueous layer is additionally extracted with three portions of ethyl acetate (100 mL each, Note 16). The combined organic layers are dried over magnesium sulfate and evaporated on a rotary evaporator. Residual ethyl acetate is azeotropically removed by adding dichloromethane (10 mL) three times and evaporating on the rotary evaporator. Trifluoroacetic acid and traces of solvent are removed under high vacuum (Note 17). The product crystallizes slowly to essentially pure material (16.9-17.1 g, 57.6-61.2 mmol, 61-65%) and can be recrystallized (diethyl ether/light petroleum 1 : 1; about 100 mL) to yield 12.1 g product (43.3 mmol, 46%, Notes 18,19).

2. Notes

1.   Boc-phenylalanine was obtained from Aldrich Chemical Co., Inc. (The submitters obtained their sample from Bachem).

2.   Tetrahydrofuran was dried over sodium/benzophenone and freshly distilled before use.

3.   Triethylamine was freshly distilled from calcium hydride.

4.   Ethyl chloroformate was freshly distilled before use.

5. A short stem, flame-polished funnel of diameter ca. 12.5 cm, free of any scratches or broken edges, was used to prevent spontaneous decomposition of diazomethane.

6.   Diazomethane was prepared by the method described (de Boer, Th. J.; Backer, H. J. *Org. Synth., Coll. Vol. IV* **1963**, 250) using a special diazomethane generator, which can be purchased from Aldrich Chemical Company, Inc. (Diazald kit Z10,025-0). The diazomethane solution was prepared by slow distillation of a reaction mixture, which was prepared by adding first a solution of 21.5 g of N-methyl-N-nitroso-p-toluenesulfonamide dissolved in 200 mL of ether to a solution of 6 g of potassium hydroxide, 10 mL of water, 35 mL of 2-(2-ethoxyethoxy)ethanol and 10 mL of ether, followed by a final addition of about 30 mL of ether until the distillate was colorless. All operations involving diazomethane

were carried out behind a blast shield and special attention should be paid to the safety instructions made in the above reference.

7. The crude diazo ketone is first obtained as a viscous yellow oil, which slowly solidifies under high vacuum. The checkers always handled the solid material behind a safety shield.

8. The crude diazo ketone (30.8-33.4 g) always contains about 10% of Boc-L-phenylalanine methyl ester formed by esterification of Boc-L-phenylalanine with diazomethane. This material can be carried through the synthesis and is removed during Step B.

9. The checkers purified the diazo ketone (1.5 g) for characterization purposes by dissolution in the minimum quantity of boiling diethyl ether (ca. 2 mL) to which was added boiling hexane (ca. 40 mL). The product does not crystallize until the solution is cooled to −20°C. The crystals are isolated (0.65 g) by filtration under vacuum, washed with hexane, and then recrystallized to give the pure diazo ketone (0.10 g). The product has the following characteristics: mp 96°C, $[\alpha]_D^{20}$ −30.4° (MeOH, c 2.57); IR (KBr) cm^{-1}: 699, 1168, 1366, 1498, 1515, 1638, 1702, 2108, 2933, 2979, 3338; ^{1}H NMR (400 MHz, CDCl$_3$) δ: 1.39 (s, 9 H, C$_4$H$_9$), 3.05 (m, 2 H, C$\underline{H}_2$Ph), 4.40 (br s, 1 H, C$\underline{H}$CH$_2$Ph), 5.07 (br s, 1 H, N$\underline{H}$), 5.20 (br s, 1 H, C$\underline{H}$N$_2$), 7.17-7.31 (m, 5 H, Ar$\underline{H}$); ^{13}C NMR (100 MHz, CDCl$_3$) δ: 29.3, 39.6, 55.5, 59.5, 81.1, 128.0, 129.6, 130.4, 137.3, 156.1, 194.3. MS (ES$^+$) m/z (rel intensity) 312.1320 [(M + Na)$^+$, calcd. for C$_{15}$H$_{19}$N$_3$O$_3$Na 312.1324], 290 [70, (M + H)$^+$]. Anal. Calcd. for C$_{15}$H$_{19}$N$_3$O$_3$: C, 62.3; H, 6.6; N, 14.5. Found: C, 62.3; H, 6.6; N, 14.1.

10. The checkers used distilled, dry tetrahydrofuran (Note 1), whereas the submitters either distilled the tetrahydrofuran without drying, or purchased a pure grade.

11. Silver trifluoroacetate was obtained from Fluka Chemika or Aldrich Chemical Company, Inc., and used as received.

12. At this stage, the material consists of large, black lumps, which should be broken up with a spatula.

13. These subsequent re-extractions are essential, since this is the most convenient method for the complete removal of the side product Boc-phenylalanine methyl ester.

14. Solid Congo Red was prepared as a well-shaken 1% w/w suspension in ethanol.

15. About 50-60 mL of hydrochloric acid are used. The color change can be obscured by the presence of the black solid, which should be allowed to settle from time to time so that the solution can be clearly viewed. The checkers observed that the pH of the aqueous phase was between 2-3 as shown by universal pH paper strips.

16. After the second extraction with ethyl acetate the pH value of the aqueous layer is shown to be pH 2-3. If necessary more hydrochloric acid is added.

17. Drying over a period of 16 hr at a pressure of 10^{-3} bar (0.75 mm) is usually sufficient.

18. The submitters obtained 17.4 g (66%). The product has the following characteristics: mp 102-103°C (the submitters obtained mp 102-106°C; Fluka catalog 1999/2000 mp 100-104°C). $[\alpha]_D^{20}$ −15.7 (MeOH, c 1.84) [Fluka catalog 1999/2000 $[\alpha]_D^{20}$ −17.5° (CH_2Cl_2, c 1.00)]; IR (KBr) cm^{-1}: 3330 (br), 2980, 1712 (br),1053; ^{1}H NMR (400 MHz, $CDCl_3$) δ: 1,40 (s, 9 H, $C_4\underline{H}_9$), 2.39-2,60 (m, 2 H, C$\underline{H}_2$Ph), 2.79-2.99 (m, 2 H, C$\underline{H}_2$COOH), 4.00-4.25 (br m, 1 H, C$\underline{H}$CH$_2$Ph), 5.02 (br s, 0.66 H, N$\underline{H}$), 5.96 (br s, 0.33 H, N$\underline{H}$), 7.10-7.35 (m, 5 H, Ar$\underline{H}$), 7.70 (br s,1 H, -CO$_2$$\underline{H}$); ^{13}C NMR (100 MHz, $CDCl_3$) δ: 28.7, 37.8, 40.6, 49.1, 80.0, 127.0, 128.9, 129.7, 138.0, 155.6, 176.8. MS (ES$^+$) m/z (rel intensity) 302.1369 [(M + Na)$^+$, calcd. for $C_{15}H_{21}NO_4Na$ 302.1368], 280 [65, (M + H)$^+$], 224 (100), 180 (55). Anal. Calcd. for $C_{15}H_{21}NO_4$: C, 64.5; H, 7.6; N, 5.0. Found C, 64.2; H, 7.6; N, 5.2. Owing to the presence of rotamers the NMR spectra measured at room temperature showed broadened or duplicated signals, and only the more intense carbon resonances have been listed. The proton and carbon spectra of the synthetic sample were identical to those of a commercial (Fluka) sample.

19. The checkers also prepared (R)-3-(tert-butyloxycarbonylamino)-4-phenylbutanoic acid from Boc-D-phenylalanine according to the same procedure. The enantiomeric purities of the (S)- and (R)-enantiomers were checked by courtesy of Mr. Eric Hortense (GlaxoSmithKline, Stevenage) separately on the corresponding methyl esters, obtained by treatment of the β-amino acids (40 mg, 0.14 mmol) with polymer-supported carbodiimide (PS-carbodiimide, Argonaut, 250 mg, 0.28 mmol) and 4-dimethylaminopyridine (8 mg, 0.07 mmol) in methanol/CH$_2$Cl$_2$ (1.4 v/v, 4 mL) for 18 hr. Subsequent filtration of the resin and purification of the crude ester by preparative reverse phase HPLC [C18 column, 10-cm x 2-cm, gradient elution, MeCN, H$_2$O, CF$_3$CO$_2$H 95:5 v/v (solvent A), H$_2$O, CF$_3$CO$_2$H 99.9:0.1 v/v (solvent B) varying from A:B 20:80 to 95:5 A:B over 20 min at a flow rate 6 mL min^{-1} afforded, after freeze-drying, the methyl ester as a colorless powder (ca. 40 mg). Upon chiral HPLC analysis on a Chiralpak AD column (25 cm, solvent EtOH/heptane 5:95 v/v , flow rate 1.0 mL min^{-1}), the (S)-enantiomer (retention time 9.9 min) exhibited an enantiomeric ratio of 99.5:0.5. The retention time of the (R)-enantiomer was 8.6 min.

Waste Disposal Information

All toxic materials were disposed of in accordance with "Prudent Practices in the Laboratory"; National Academy Press; Washington, DC, 1995.

3. Discussion

β-Amino acids are useful precursors for the construction of β-peptides,[2,3] α-substituted β-amino acids[4] and related compounds.[5] They can be prepared enantiomerically pure by homologation of α-amino acids using the Arndt-Eistert method.

160

The suitably protected amino acid is activated as the mixed anhydride and treated with diazomethane to produce the corresponding diazo ketone. Rearrangement in the presence of water furnishes the β-amino acid. Diazomethane contains varying amounts of water, which is able to hydrolyze the activated amino acid. This leads to subsequent methylation by diazomethane to form the methyl ester as a side product. This cannot easily be removed from the diazo ketone, but can be separated during work-up of the homologated amino acids.

Substitution of diazomethane by the less hazardous trimethylsilyl-substituted diazomethane (TMS-CHN$_2$)[6] is not possible, since TMS-CHN$_2$ is not acylated by mixed anhydrides.

The diazo ketones that are synthesized as intermediates are not only useful for the preparation of β-amino acids but may serve as versatile starting materials in different reactions,[7] e.g. preparation of 3-azetidinones[8] or 2-aminocyclopentanones.[9]

The procedure described here has been used for the synthesis of further Boc-protected β-amino acids:

Synthesis of Boc-Protected β-Amino Acids

entry	product	yield(%)	entry	product	yield(%)
1	Boc-NH-CH(CH$_3$)-CH$_2$-CO$_2$H	58	6	Boc-NH-CH(CH$_2$SMe)-CH$_2$-CO$_2$H	58
2	Boc-NH-CH(CH$_2$CH(CH$_3$)$_2$)-CH$_2$-CO$_2$H	36	7	Boc-NH-CH(CH$_2$CH$_2$NHCbz)-CH$_2$-CO$_2$H	32
3	proline Boc-CO$_2$H	80	8	Boc-NH-CH(CH$_2$CO$_2$Me)-CH$_2$-CO$_2$H	74
4	Boc-NH-CH(CH$_2$-cyclohexyl)-CH$_2$-CO$_2$H	58	9	Boc-NH-CH(CH$_2$-indolyl)-CH$_2$-CO$_2$H	58
5	Boc-NH-CH(CH$_2$OBn)-CH$_2$-CO$_2$H	44			

1. Institut für Organische Chemie, Universität Stuttgart, Pfaffenwaldring 55, D-70569 Stuttgart, Germany.

2. Seebach, D.; Matthews, J. L. *J. Chem. Soc., Chem. Commun.* **1997**, 2015-2022; Matthews, J. L.; Braun, C.; Guibourdenche, C.; Overhand, M.; Seebach, D. In "Enantioselective Synthesis of β-Amino Acids", Juaristi, E., Ed., Wiley-VCH: New York, **1997**; pp. 105-126; Matthews, J. L.; Overhand, M.; Kühnle, F. N. M.; Ciceri, P. E.; Seebach, D. *Liebigs Ann./Recl* .**1997**, 1371-1379; Seebach, D.; Matthews, J. L.; Meden, A.; Wessels, T.; Baerlocher, C.; McCusker, L. B. *Helv. Chim. Acta* **1997**, *80*, 173-182; Seebach, D.; Abele, S.;

Sifferlen, T.; Hänggi, M.; Gruner, S.; Seiler, P. *Helv. Chim. Acta* **1998**, *81*, 2218-2243; Abele, S.; Guichard, G.; Seebach, D. *Helv. Chim. Acta* **1998**, *81*, 2141-2156; Seebach, D.; Abele, S.; Gademann, K.; Guichard, G.; Hintermann, T.; Jaun, B.; Matthews, J. L.; Schreiber, J. V.; Oberer, L.; Hommel, U.; Widmer, H. *Helv. Chim. Acta* **1998**, *81*, 932-982; Seebach, D.; Abele, S.; Schreiber, J. V.; Martinoni, B.; Nussbaum, A. K.; Schild, H.; Schulz, H.; Hennecke, H.; Woessner, R.; Bitsch, F. *Chimia* **1998**, *52*, 734-739; Matthews, J. L.; Gademann, K.; Jaun, B.; Seebach, D. *J. Chem. Soc., Perkin Trans. 1* **1998**, 3331-3340; Seebach, D.; Schreiber, J. V.; Arvidsson, P. I.; Frackenpohl, J. *Helv. Chim. Acta* **2001**, *84*, 271-279.

3. Appella, D. H.; Christianson, L. A.; Karle, I. L.; Powell, D. R.; Gellman, S. H. *J. Am. Chem. Soc.* **1996**, *118*, 13071-13072; Appella, D. H.; Christianson, L. A.; Klein, D. A.; Powell, D. R.; Huang, X.; Barchi, J. J.; Jr.; Gellman, S. H. *Nature* **1997**, *387*, 381-384; Krauthäuser, S.; Christianson, L. A.; Powell, D. R.; Gellman, S. H. *J. Am. Chem. Soc.* **1997**, *119*, 11719-11720; Chung, Y. J.; Christianson, L. A.; Stanger, H. E.; Powell, D. R.; Gellman, S. H. *J. Am. Chem. Soc.* **1998**, *120*, 10555-10556; Gellman, S. H. *Acc. Chem. Res.* **1998**, *31*, 173-180; Huck, B. R.; Fisk, J. D.; Gellman, S. H. *Org. Lett.* **2000**, *2*, 2607-2610.

4. Podlech, J.; Seebach, D. *Liebigs Ann.* **1995**, 1217-1228.

5. Podlech, J.; Seebach, D. *Angew. Chem.* **1995**, *107*, 507-509; *Angew. Chem. Int. Ed. Engl.* **1995**, *34*, 471-472; Guibourdenche, C.; Seebach, D.; Natt, F. *Helv. Chim. Acta* **1997**, *80*, 1-13; Guibourdenche, C.; Podlech, J.; Seebach, D. *Liebigs Ann.* **1996**, 1121-1129; Limal, D.; Semetey, V.; Dalbon, P.; Jolivet, M.; Briand, J.-P. *Tetrahedron Lett.* **1999**, *40*, 2749-2752.

6. Podlech, J. *J. Prakt. Chem. Chem.-Ztg.* **1998**, *340*, 679-682.

7. Ye, T.; McKervey, M. A. *Chem. Rev.* **1994**, *94*, 1091-1160; Doyle, M. P.; McKervey, M. A. *J. Chem. Soc., Chem. Commun.* **1997**, 983-989.

8. Podlech, J.; Seebach, D. *Helv. Chim. Acta* **1995**, *78*, 1238-1246.

9. Sengupta, S.; Das, D. *Synth. Commun.* **1998**, *28*, 403-408.

Appendix

Chemical Abstracts Nomenclature (Collective Index Number);
(Registry Number)

(S)-3-(tert-Butyloxycarbonylamino)-4-phenylbutanoic acid: Benzenebutanoic acid,

β-[[(1,1-dimethylethoxy)carbonyl]amino]-, (S)- (9); (51871-62-6)

Diazomethane: Methane, diazo- (8,9); (334-88-3)

(S)-3-(tert-Butyloxycarbonylamino)-1-diazo-4-phenylbutan-2-one: Carbamic acid,

[3-diazo-2-oxo-1-(phenylmethyl)propyl]-, 1,1-dimethylethyl ester, (S)- (9); (60398-41-6)

Boc-Phenylalanine: L-Phenylalanine, N-[(1,1-dimethylethoxy)carbonyl]- (9); (13734-34-4)

Triethylamine (8); Ethanamine, N,N-diethyl- (9); (121-44-8)

Ethyl chloroformate: Formic acid, chloro-, ethyl ester (8); Carbonochloridic acid, ethyl ester

(9); (541-41-3)

Silver trifluoroacetate: Acetic acid, trifluoro-, silver(1+) salt (8,9); (2966-50-9)

Trifluoroacetic acid: Acetic acid, trifluoro- (8, 9); (76-05-1)

PHOTO-INDUCED RING EXPANSION OF 1-TRIISOPROPYLSILYLOXY-1-AZIDOCYCLOHEXANE: PREPARATION OF ε-CAPROLACTAM

[2H-Azepin-2-one, hexahydro- from Silane, [(1-azidocyclohexyl)oxy]tris(1-methylethyl)-]

Submitted by Jade D. Nelson, Dilip P. Modi, and P. Andrew Evans.[1]

Checked by Patrick Foyle, Peter Belica, and Steven Wolff.

1. Procedure

A. 1-Triisopropylsilyloxy-1-azidocyclohexane: A 2-L, two-necked, round-bottomed flask is equipped with a magnetic stirrer, argon inlet, and a rubber septum (Note 1). The flask is charged with freshly-distilled 1-triisopropylsilyloxycyclohexene (25.47 g, 100 mmol, Note 2) and anhydrous dichloromethane (1.0 L, Note 3). Azidotrimethylsilane (68.5 mL, 500 mmol, Note 4) is added via syringe, immediately followed by anhydrous Dowex® 50 x 8-100 (24.98 g, Note 5) in a single portion from a dry flask. The suspension is stirred vigorously at ambient temperature for ca. 48 hr (Note 6). The reaction mixture is filtered to recover the Dowex® resin and solvent is removed under reduced pressure to afford a clear, colorless oil. The crude oil is applied to a 5 x 13-cm column of silica gel (120 g, 230-400 mesh packed with hexanes). The column is quickly eluted with hexanes (750 mL) and fractions are collected in 25-mL test tubes (Note 7). The fractions containing the desired product are identified by thin layer chromatography, combined, and concentrated under reduced pressure to afford the 1-triisopropylsilyloxy-1-azidocyclohexane (26.75 g, 90%) as a colorless oil (Note 8).

B. *ε-Caprolactam*: The reaction apparatus (see Fig. 1, Note 9) is charged with 1-triisopropylsilyloxy-1-azidocyclohexane (25.9 g, 87.1 mmol) and cyclohexane (435 mL, Note 10). The clear, colorless solution is purged with nitrogen for a 15 min period to exclude oxygen. The solution is cooled to ca. 0°C and irradiated ($\geq$200 nm) for 3.5 hr (Note 11). The solvent is removed under reduced pressure to afford a pale yellow oil. The crude oil is dissolved in dichloromethane (ca. 10 mL) and applied to a 6 x 6-cm column of silica gel (88 g, 230-400 mesh packed with 1 : 1 ethyl acetate/hexanes). An additional 1 cm of silica gel is added to the top of the column and the mixture is stirred with a glass rod to homogenize the layer, rinsed with eluent, and the column is repacked. The column is eluted with 1 : 1 ethyl acetate/hexanes (200 mL, Note 12), then with 1 : 9 methanol/ethyl acetate (500 mL), and fractions are collected in 25-mL test tubes. The fractions containing the desired product are identified by thin layer chromatography (Note 13), combined, and concentrated under reduced pressure to afford *ε-caprolactam* (8.20 g, 83%, Note 14) as an off-white solid.

2. Notes

1.   The assembled glassware was flame-dried under high vacuum, then cooled to ambient temperature under a positive pressure of dry argon.

2.   The following procedure was used to prepare 1-triisopropylsilyloxy-cyclohexene. A 500-mL, three-necked, round-bottomed flask was charged with 9.81 g

166

(0.10 mol) of cyclohexanone, 180 mL of dry dichloromethane (anhydrous grade from Aldrich Chemical Company, Inc.), and 20.9 mL (0.15 mol) of triethylamine. The mixture was cooled to −20° and a solution of 32.25 mL of triisopropylsilyl trifluoromethanesulfonate[2] in 20 mL of dichloromethane was added dropwise, while maintaining the temperature of the reaction below 5°C. After completion of the addition, the reaction mixture was stirred at 0-5°C for 1 hr, then at ambient temperature for 2 hr. The reaction mixture was washed with 80 mL of brine, then dried with magnesium sulfate. After filtration to remove the drying agent, the filtrate was concentrated under vacuum to yield a residue consisting of two phases. The residue was diluted with 200 mL of hexanes, washed with 50 mL of brine, dried with magnesium sulfate, and filtered. The filtrate was concentrated under vacuum to give 28.37 g of a pale yellow oil. Distillation afforded 24.97 g of a colorless oil, bp 75-85°C (0.2-0.3 torr, 0.15-0.23 mm). The NMR spectrum of this material indicated the possible presence of cyclohexanone. Redistillation gave 20.25 g (78%) of 1-triisopropylsilyloxycyclohexene as a colorless oil, bp 85-90°C (0.3 torr, 0.23 mm).

3. Dichloromethane was distilled from calcium hydride under an atmosphere of nitrogen immediately prior to use.

4. The azidotrimethylsilane was purchased from Acros Organics and used without further purification.

5. Dowex[®] 50X8-100 was purchased from Acros Organics and dried in the following manner. Approximately 50 g of the commercially available resin is washed with anhydrous methanol (3 x 50 mL), then with anhydrous ethyl ether (2 x 50 mL) in a Buchner funnel. The granular solid is dried under high vacuum for ca. 24 hr.

6. The progress of the reaction is monitored by thin layer chromatography, eluting with hexanes, then dried quickly with a stream of nitrogen and eluted a second time with hexanes (product R_f = 0.75, visualized and developed using UV light and $KMnO_4$, respectively). The checkers found that the reaction was not complete after 72 hr. Stirring was continued until TLC analysis indicated the absence of the enol ether (5-6 days).

7.   Discoloration of the column is often observed, which may be due to hydrazoic acid formation upon hydrolysis of the azidotrimethylsilane.

8.   The product exhibits the following spectroscopic and analytical properties: IR (neat) cm^{-1}: 2948, 2892, 2865, 2104; ^{1}H NMR (250 MHz, C$_6$D$_6$) δ: 1.04-1.18 (m, 21 H), 1.26-1.57 (m, 8 H), 1.66-1.76 (m, 2 H); ^{13}C NMR (62.5 MHz, C$_6$D$_6$) δ: 13.56, 18.45, 23.4, 25.16, 38.50, 91.57; HRMS (EI) calcd for C$_{11}$H$_{24}$N$_3$OSi 254.1689 found 254.1674. Anal. calcd. for C$_{15}$H$_{31}$N$_3$OSi: C, 60.56; H, 10.50; N, 14.12 Found: C, 60.51; H, 10.76; N, 14.04.

9.   The reaction apparatus (Figure 1) requires a tubular quartz flask with two side arms and a large ground-glass joint at the top to accommodate the water-cooled UV lamp. One of the side arms is fitted with a Teflon® tube, the second side arm is fitted with a rubber septum and wide-bore needle, which has a Teflon® tube leading to a nitrogen bubbler. A vigorous stream of nitrogen is allowed to flow into the flask through the Teflon® tube to agitate the solution during irradiation. The entire apparatus is placed inside a vacuum-jacketed Dewar flask filled with ice to maintain the reaction temperature at ca. 0°C.

10. Cyclohexane was purchased from Acros Organics and used without further purification.

11. The UV lamp generates a great deal of heat, and thus it is necessary to interrupt the reaction *ca.* every hour to remove water from the Dewar and replenish it with ice.

12. The triisopropylsilanol side-product is eluted first through the column. Alternatively, the bulk of this material may be removed via distillation under reduced pressure (50-60°C at 2-3 mmg).

13. The progress of the reaction is monitored by thin layer chromatography, eluting with ethyl acetate (product R$_f$ = 0.06, visualized and developed using UV light and KMnO$_4$, respectively).

14. The product exhibits the following spectroscopic and analytical properties: mp 68-69°C; lit.[3] 68-70°C; IR (CHCl$_3$) cm^{-1}: 3420, 3293, 3224, 3019, 2937, 2859, 1660; ^{1}H

NMR (250 MHz, C_6D_6) δ: 1.14-1.33 (m, 6 H), 2.21-2.25 (m, 2 H), 2.68-2.74 (m, 2 H), 8.33 (bs, 1 H); HRMS (EI) calcd for $C_6H_{11}NO$: 113.0841; found: 113.0833.

Waste Disposal Information

All toxic materials were disposed of in accordance with "Prudent Practices in the Laboratory"; National Academy Press; Washington, DC, 1995.

3. Discussion

The synthesis of lactams has attracted considerable attention in recent years. This is presumably because they represent versatile synthetic intermediates that are present in many biologically important molecules.[4] Despite the wide range of methodologies that have been examined for the synthesis of lactams,[5,6] the Beckmann[7] and Schmidt[8] rearrangements still remain by far the most convenient and general methods. The strongly acidic conditions required for the Schmidt rearrangement often lead to undesired by-products. This is a major limitation particularly with acid-labile substrates.

The method outlined here[9] represents a convenient and environmentally benign Schmidt rearrangement, in which the azidohydrin is prepared using a recyclable acid catalyst and trimethylsilyl azide, a non-explosive source of azide.[10] Photolysis of the azidocyclohexane results in the ring expansion, probably through the formation of a reactive nitrene. The by-products from this reaction are gases or innocuous silanes. The main limitation with the method is that at present the ring expansion is not regioselective, as exemplified by entries 1 and 2 in the Table, in which a mixture of regioisomers is obtained.

A further advantage of this protocol is that it allows the azidohydrin intermediate to be isolated. This will facilitate important mechanistic work to clarify the nature of the reactive species responsible for the ring expansion. Although only the preparation of azepin-2-

ones have been reported, other ring sizes have also been successfully examined. Hence, this method provides a general method for the preparation of lactams.

1. Brown Laboratory, Department of Chemistry and Biochemistry, University of Delaware, Newark, Delaware 19716. Present address: Department of Chemistry, Indiana University, Bloomington, IN 47405.

2. Corey, E. J.; Cho, H.; Rücker, C.; Hua, D.H. *Tetrahedron Lett.* **1981**, *22*, 3455.

3. Marvel, C. S.; Eck, J. C. *Org. Synth., Coll. Vol. II* **1943**, 371.

4. Evans, P. A.; Holmes, A. B. *Tetrahedron* **1991**, *47*, 9131.

5. (a) Evans, P. A.; Holmes, A. B.; Russell, K. *Tetrahedron: Asymmetry* **1990**, *1*, 593; (b) Evans, P. A.; Holmes, A. B.; Russell, K. *Tetrahedron Lett.* **1992**, *33*, 6857; (c) Evans, P. A.; Holmes, A. B.; Russell, K. *J. Chem. Soc., Perkin Trans. I* **1994**, 3397; (d) Evans, P. A.; Holmes, A. B.; McGeary, R. P.; Nadin, A.; Russell, K.; O'Hanlon, P. J.; Pearson, N. D. *J. Chem. Soc. Perkin Trans. I* **1996**, 123.

6. (a) Kawase, M. *J. Chem. Soc., Chem. Commun.* **1992**, 1076; (b) Vedejs, E.; Sano, H. *Tetrahedron Lett.* **1992**, *33*, 3261; (c) Kim, S.; Joe, G. - H.; Do, J. - Y. *J. Am. Chem. Soc.* **1993**, *115*, 3328; (d) McGee, D. I.; Ramaseshan, M. *Synlett*, **1994**, 743; (e) Suda, K.; Sashima, M.; Izutsu, M.; Hino, F. *J. Chem. Soc., Chem. Commun.* **1994**, 949; (f) Robl, J. A.; Cimarusti, M. P.; Simpkins, L. M.; Weller, H. N.; Pan, Y. Y.; Malley, M.; Dimarco, J. D. *J. Am. Chem. Soc.*, **1994**, *116*, 2348; (g) Coates, B.; Montgomery, D.; Stevenson, P. J. *Tetrahedron* **1994**, *50*, 4025; (h) Nadin, A.; Derrer, S.; McGeary, R. P.; Goodman, J. M.; Raithby, P. R.; Holmes, A. B. *J. Am. Chem. Soc.* **1995**, *117*, 9768. (i) Milligan, G. L.; Mossman, C. J.; Aubé, J. *J. Am. Chem. Soc.* **1995**, *117*, 10449. (j) Barluenga, J.; Tomás, M.; Ballesteros, A.; Santamaria, J.; Suárez-Sobrino, A. *J. Org. Chem.* **1997**, *62*, 9229. (k) Benati, L.; Nanni, D.;

Sangiorgi, C.; Spagnolo, P. *J. Org. Chem.* **1999**, *64*, 7836. (l) Bottcher, G.; Reissig, H.- U. *Synlett* **2000**, 725 and pertinent references cited therein.

7. (a) Schmidt, K. F. *Z. Angew. Chem.* **1923**, *36*, 511; (b) Wolff, H. *Org. React.* **1946**, *3*, 307.

8. Beckmann, E. *Chem. Ber.* **1886**, *19*, 988.

9. Evans, P. A.; Modi, D. P. *J. Org. Chem.* **1995**, *60*, 6662.

10. Trimethylsilyl azide has been reported to be a non-explosive alternative to hydrazoic acid: Birkofer, L.; Wegner, P. *Org. Synth., Coll. Vol. VI* **1988**, 1030.

TABLE[9]

PREPARATION OF AZEPIN-2-ONES VIA THE PHOTO-INDUCED RING EXPANSION OF AZIDOCYCLOHEXANES

Entry	α-Azidohydrin[a]	Azepin-2-ones		Yield (%)[b]

Entry	α-Azidohydrin[a]	Azepin-2-ones	Yield (%)[b]
1	(structure: N$_3$, OSi-iso-Pr$_3$, Me)	(structure) R_1 = Me; R_2 = H / R_1 = H; R_2 = Me	75[c]
2	(structure: N$_3$, OSi-iso-Pr$_3$, Me)	(structure) R_1 = Me; R_2 = H / R_1 = H; R_2 = Me	87[d]
3	(structure: N$_3$, OSi-iso-Pr$_3$, R)	(structure) R = Me / R = tert-Bu	89 / 85
4	(structure: N$_3$, OSi-iso-Pr$_3$, R R)	(structure) R = R = Me / R=R= O(CH$_2$)$_2$O	85 / 82

5	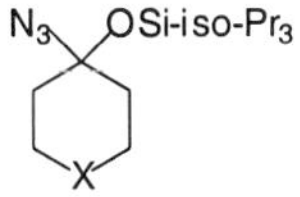	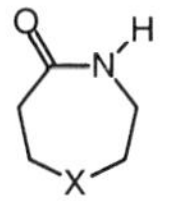	X = O	83
			X = S	64

[a]Photolysis with ultraviolet light at 0°C on a 0.25-mmol reaction scale for *ca.* 1 hr.

[b]Isolated yields. [c]1.6 : 1 Mixture regioisomers. [d]1 : 1 Mixture of regioisomers.

Figure 1

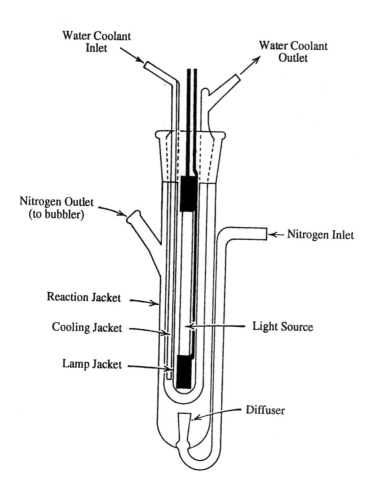

Water Coolant Inlet

Water Coolant Outlet

Nitrogen Outlet (to bubbler)

Nitrogen Inlet

Reaction Jacket

Cooling Jacket

Light Source

Lamp Jacket

Diffuser

Appendix

Chemical Abstracts Nomenclature (Collective Index Number); (Registry Number)

1-Triisopropylsilyloxy-1-azidocyclohexane: Silane, [(1-azidocyclohexyl)oxy]tris(1-methylethyl)- (13); (172090-42-5)

1-Triisopropylsilyloxycyclohexene: Silane, (1-cyclohexen-1-yloxy)tris(1-methylethyl)- (11); (80522-46-9)

ε-Caprolactam: 2H-Azepin-2-one, hexahydro- (8, 9); (105-60-2)

Azidotrimethylsilane: HIGHLY TOXIC: Silane, azidotrimethyl- (8, 9); (4648-54-8)

Cyclohexanone (8, 9); (108-94-1)

Triethylamine (8); Ethanamine, N, N-diethyl- (9); (121-44-8)

Triisopropylsilyl trifluoromethanesulfonate: Methanesulfonic acid, trifluoro-, tris(1-methylethyl)silyl ester (11); (80522-42-5)

(3,4,5-TRIFLUOROPHENYL)BORONIC ACID-CATALYZED AMIDE FORMATION FROM CARBOXYLIC ACIDS AND AMINES: N-BENZYL-4-PHENYLBUTYRAMIDE

[Benzenebutanamide, N-(phenylmethyl)-]

Submitted by Kazuaki Ishihara,[1] Suguru Ohara,[2] and Hisashi Yamamoto.[2]

Checked by David T. Amos and Rick L. Danheiser.

1. Procedure

A. (3,4,5-Trifluorophenyl)boronic acid. A 500-mL, three-necked, round-bottomed flask containing magnesium turnings (1.94 g, 80 mmol) is equipped with a rubber septum, a 20-mL pressure-equalizing dropping funnel fitted with a rubber septum, a Teflon-coated magnetic stirring bar, and a reflux condenser fitted with an argon inlet adapter. The system is flame-dried and flushed with argon. Anhydrous ether (200 mL, Note 1) is introduced to cover the magnesium, a crystal of iodine is added, and the mixture is heated to reflux in an oil bath. The dropping funnel is filled with 1-bromo-3,4,5-trifluorobenzene (8.36 mL, 14.8

176

g, 70.0 mmol, Note 2) and ca. 1 mL is added to the boiling reaction mixture. After reaction has commenced, the oil bath is removed, and the remainder of the aryl bromide is added slowly at a rate sufficient to maintain reflux (addition time ca. 1 hr). The resulting mixture is stirred for an additional 2 hr. During this period, a flame-dried, 500-mL, single-necked, round-bottomed flask equipped with a Teflon-coated magnetic stirring bar, a rubber septum, and an argon inlet is charged with dry tetrahydrofuran (THF, 50 mL, Note 3) and trimethyl borate (15.7 mL, 14.5 g, 140 mmol, Note 4). The mixture is cooled to 0°C, and the ether solution of (3,4,5-trifluorophenyl)magnesium bromide prepared above is introduced in one portion via a double-ended needle. The reaction mixture is allowed to warm to room temperature, stirred for 1 hr, and then treated with 200 mL of saturated ammonium chloride solution. The organic layer is separated and the aqueous layer is extracted with three 100-mL portions of ethyl acetate. The combined organic layers are washed with brine (100 mL), dried over anhydrous magnesium sulfate, filtered, and concentrated under reduced pressure. The resulting white solid is dissolved in a minimal amount of hot (65°C) ethyl acetate, allowed to cool to room temperature, and then 600 mL of hexane is added. The resulting solution is allowed to stand overnight and then filtered to afford pure (3,4,5-trifluorophenyl)boronic acid as white crystals. Further recrystallization of the mother liquor 3-4 times provides a total of 6.3 g (51%) of (3,4,5-trifluorophenyl)boronic acid (Notes 5 and 6).

B. *N-Benzyl-4-phenylbutyramide*. A flame-dried, 200-mL, single-necked, round-bottomed flask is equipped with a Teflon-coated magnetic stirring bar and a Soxhlet extractor containing a thimble filled with 3 g of calcium hydride and topped with a reflux condenser fitted with an argon inlet (Note 7). The reaction flask is charged with 4-phenylbutyric acid (5.42 g, 33.0 mmol, Note 8), benzylamine (3.28 mL, 30.0 mmol, Note 9), and (3,4,5-trifluorophenyl)boronic acid (52.8 mg, 0.300 mmol) in toluene (60 mL) and then heated in an oil bath. The reaction mixture is brought to reflux (bath temperature 120°C), and after 16 hr is cooled to ambient temperature and diluted with 80 mL of

177

dichloromethane. The organic layer (Note 10) is washed with 1.0 M hydrochloric acid (HCl, 100 mL) and brine (100 mL), dried over anhydrous magnesium sulfate, filtered, and concentrated under reduced pressure. The resulting yellow residue is recrystallized from ethyl acetate and hexane to provide pure N-benzyl-4-phenylbutyramide (ca. 6-7 g) as white crystals. The mother liquor is concentrated and the residue is purified by flash chromatography on silica gel (Note 11) to provide additional product as a white solid. The total combined yield of N-benzyl-4-phenylbutyramide is 7.11-7.18 g (94-95%, Note 12).

2. Notes

1. Ethyl ether was distilled from sodium-benzophenone ketyl before use.

2. 1-Bromo-3,4,5-trifluorobenzene was purchased from Aldrich Chemical Company, Inc., and used without further purification.

3. Tetrahydrofuran was distilled from sodium-benzophenone ketyl before use.

4. Trimethyl borate was purchased from Tokyo Kasei Kogyo Co., Ltd. or Aldrich Chemical Company, Inc. and used without further purification.

5. The product consists of a mixture of (3,4,5-trifluorophenyl)boronic acid and varying amounts of (3,4,5-trifluorophenyl)boronic anhydride.

6. The submitters obtained the product in 89% yield. (3,4,5-Trifluorophenyl)boronic acid has the following physical properties: TLC R_f = 0.63 (10:1 ethyl acetate/methanol); mp 249-252°C, IR (KBr) cm^{-1}: 3077, 2359, 1616, 1530, 1217, 1038; ^{1}H NMR (300 MHz, CDCl$_3$) δ: 4.74-4.82 [br, 0.28 H (for monomer)], 7.35 (t, 0.28 H, J = 7.0 (for monomer)), 7.77 [t, 1.72 H, J = 7.9 (for trimer)]; ^{13}C NMR (125 MHz, CD$_3$OD) δ: 118.6 (dd, J = 4.6, 15.0), 130.5-132.6 (br m), 142.2 (dt, J = 249.8, 15.1), 152.2 (ddd, J = 249.7, 9.4, 2.3). Anal. Calcd for (C$_6$H$_2$OBF$_3$)$_3$: C, 45.64; H, 1.28. Found: C, 45.32; H, 1.64 (microanalysis was carried out on a sample that was dried at 60-80°C under high vacuum for 2 hr).

7. The submitters used a 10-mL, pressure-equalized addition funnel [containing a cotton plug, calcium hydride (ca. 3 g, lumps), and sea sand (ca. 1 g)] in place of the Soxhlet extractor. The submitters employed calcium hydride (ca. 1-10 mm, No. 068-34) purchased from Nacalai Tesque, Inc. Alternatively, 4Å molecular sieves can be used in place of calcium hydride.

8. 4-Phenylbutyric acid (>99%) was purchased from Tokyo Kasei Kogyo Co., Ltd. or Aldrich Chemical Company, Inc., and used without further purification.

9. The submitters purchased benzylamine (99%) from Nacalai Tesque, Inc. and used it without further purification. The checkers obtained the amine from Aldrich Chemical Company, Inc., and distilled it from calcium hydride.

10. The submitters report obtaining a two-phase mixture upon cooling and adding dichloromethane. The organic layer was separated and the aqueous layer was extracted with dichloromethane (80 mL), and treated with 1.0 M aqueous sodium hydroxide solution (100 mL). The combined organic layers were then washed with HCl and brine as described in the procedure.

11. Chromatography was performed using a 3-cm x 10-cm column packed with 35 g of silica gel (230-400 mesh, No. 9385) purchased from E. Merck Co. The product was eluted with 100 mL of 25% and 200 mL of 33% ethyl acetate-hexane. The checkers observed that N-benzyl-4-phenylbutyramide has a TLC R_f value of 0.4 in 50% ethyl acetate-hexane.

12. N-Benzyl-4-phenylbutyramide has the following physical properties: mp 79-80°C; IR (CH_2Cl_2) cm^{-1}: 1671, 1510, 1460, 1271, 1260; ^{1}H NMR (300 MHz, $CDCl_3$) δ: 1.96-2.06 (m, 2 H), 2.22 (t, 2 H, J = 6.9), 2.67 (t, 2 H, J = 8.0), 4.44 (d, 2 H, J = 6.0), 5.62 (br, 1 H), 7.15-7.34 (m, 10 H); ^{13}C NMR (75.4 MHz, CD_3OD) δ: 28.8, 36.2, 36.4, 44.0, 126.9, 128.1, 128.5, 129.3, 129.4, 129.5, 140.0, 142.8, 175.6 (C=O). Anal. Calcd for $C_{17}H_{19}NO$: C, 80.60; H, 7.56; N, 5.53. Found: C, 80.34; H. 7.67; N, 5.58.

179

Waste Disposal Information

All toxic materials were disposed of in accordance with "Prudent Practices in the Laboratory"; National Academy Press; Washington, DC, 1995.

3. Discussion

There are several different routes to carboxamides.[3] In most of these reactions, a carboxylic acid is converted to a more reactive intermediate, e.g. the acid chloride, which is then allowed to react with an amine. For practical reasons, it is preferable to form the reactive intermediate in situ.[4] Arylboronic acids with electron-withdrawing groups such as (3,4,5-trifluorophenyl)boronic acid act as highly efficient catalysts in the amidation between carboxylic acids and amines.[5] (3-Nitrophenyl)boronic acid and [3,5-bis(trifluoromethyl)phenyl]boronic acid are also effective amidation catalysts and commercially available.

Acyloxyboron intermediates generated from carboxylic acids and boron reagents such as BR_3 (R=C_8H_{17}, OMe),[6a] $ClB(OMe)_2$,[6a] $HB(OR)_2$ (R=i-Pr, t-Am),[6a] $BH_3 \cdot Et_3N$ (R=Me, Bu),[6b] $BF_3 \cdot Et_2O$[6c] and catecholborane[6d] react with amines to furnish amides in moderate to good yield, but only in uniformly stoichiometric reactions. In these amidations, boron reagents transform into inactive boron species after the reaction of acyloxyboron derivatives and amines. However, arylboronic acids with electron-withdrawing substituents at the aryl group can be used to circumvent these difficulties, since they are water-, acid-, and base-tolerant Lewis acids that can generate acyloxyboron species. Their strong Lewis acidity enhances the rate of the generation of acryloxyboron species and their reactivity with amines.

To indicate the generality and scope of (3,4,5-trifluorophenyl)boronic acid-catalyzed amidation, the reaction is examined with various structurally diverse carboxylic acids and

180

primary or secondary amines (Table I). In most cases, the reactions proceed cleanly, and the desired carboxylic amides are obtained in high yields. The catalyst is useful for effecting reaction not only of primary but also of secondary amines with various carboxylic acids. Sterically-hindered 1-adamantanecarboxyic acid is easily amidated at reflux in mesitylene. Aromatic substrates such as anilines and benzoic acid also react well under similar conditions. The catalytic amidation of optically active aliphatic α-hydroxycarboxylic acids with benzylamine proceeds with no measurable loss (<2%) of enantiomeric purity under conditions of reflux in toluene. However, slight racemization is observed in the case of (S)-(+)-mandelic acid.

In addition, lactams can be prepared by the present technique under heterogeneous conditions although most amino acids are barely soluble in nonaqueous solvents (Table II). Interestingly, (S)-(-)-proline selectively gives the cyclic dimer with no measurable loss of enantiomeric purity.

The proposed mechanism of the boron-catalyzed amidation is depicted in the Figure.[5] It has been ascertained by ^{1}H NMR analysis that monoacyloxyboronic acid 1 is produced by heating the 2:1 mixture of 4-phenylbutyric acid and [3,5-bis(trifluoromethyl)phenyl]boronic acid in toluene under reflux with removal of water. The corresponding diacyloxyboron derivative is not observed at all. When 1 equiv of benzylamine is added to a solution of 1 in toluene, the amidation proceeds even at room temperature, but the reaction stops before 50% conversion because of hydrolysis of 1. These experimental results suggest that the rate-determining step is the generation of 1.

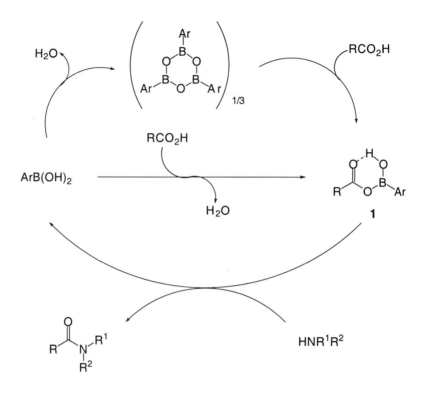

Figure. Proposed Catalytic Cycle

1. Research Center for Advanced Waste and Emission Management (ResCWE), Nagoya University, Furo-cho, Chikusa, Nagoya 464-8603, Japan.

2. School of Engineering, Nagoya University, Furo-cho, Chikusa, Nagoya 464-8603, Japan.

3. For a review of the synthesis of amides and related compounds, see: Benz, G. In "Comprehensive Organic Synthesis"; Trost, B. M.; Fleming, I.; Winterfeldt, E., Eds.; Pergamon Press: New York, 1991; Vol, 6, Chapter 2.3, p. 381.

4. For catalytic amidations between carboxylic acids and amines, see: TiCl$_4$ (a) Nordahl, Å.; Carlson, R. Acta Chem. Scand., Ser. B. 1988, 28; Ti(Oi-Pr)$_4$ (b)

Mader, M.; Helquist, P. *Tetrahedron Lett.* **1988**, *29*, 3049; Ph_3SbO/P_4S_{10} (c) Nomura, R.; Nakano, T.; Yamada, Y.; Matsuda, H. *J. Org. Chem.* **1991**, *56*, 4076; $Sb(OEt)_3$ (d) Ishihara, K.; Kuroki, Y.; Hanaki, N.; Ohara, S.; Yamamoto, H. *J. Am. Chem. Soc.* **1996**, *118*, 1569.

5.   (a) Ishihara, K.; Ohara, S.; Yamamoto, H. *J. Org. Chem.* **1996**, *61*, 4196; (b) Ishihara, K.; Ohara, S.; Yamamoto, H. *Macromolecules* **2000**, *33*, 3511; (c) Ishihara, K.; Kondo, S.; Yamamoto, H. *Synlett* **2001**, in press.

6.   (a) Pelter, A.; Levitt, T. E.; Nelson, P. *Tetrahedron* **1970**, *26*, 1539; (b) Trapani, G.; Reho, A.; Latrofa, A. *Synthesis* **1983**, 1013; (c) Tani, J.; Oine, T.; Inoue, I. *Synthesis* **1975**, 714; (d) Collum, D. B.; Chen, S.-C.; Ganem, B. *J. Org. Chem.* **1978**, *43*, 4393.

Appendix
Chemical Abstracts Nomenclature (Collective Index Number);
(Registry Number)

(3,4,5-Trifluorophenyl)boronic acid: Boronic acid, (3,4,5-trifluorophenyl)- (13); (143418-49-9)

N-Benzyl-4-phenylbutyramide: Benzenebutanamide, N-(phenylmethyl)- (13); (179923-27-4)

Magnesium (8,9); (7439-95-4)

Iodine (8,9); (7553-56-2)

1-Bromo-3,4,5-trifluorobenzene: Benzene, 5-bromo-1,2,3-trifluoro- (13); (138526-69-9)

Trimethyl borate: Boric acid, trimethyl ester (8,9); (121-43-7)

4-Phenylbutyric acid: Butyric acid, 4-phenyl- (8); Benzenebutanoic acid (9); (1821-12-1)

Benzylamine (8); Benzenemethanamine (9); (100-46-9)

TABLE 1
EXAMPLES OF AMIDATION CONDENSATION BETWEEN CARBOXYLIC
ACIDS AND AMINES CATALYZED BY (3,4,5-TRIFLUOROPHENYL)BORONIC ACID[a]

Carboxylic Acid	Amine	Solvent	Time (hr)	Yeild (%)
Ph~~CO$_2$H (butyl chain)	HN (3,5-dimethylpiperidine)	Toluene	48[b]	96[b]
Ph~~CO$_2$H (propyl chain)	Bu$_2$NH	Mesitylene	14.5	99
Ph~~CO$_2$H (propyl chain)	PhNH$_2$	Mesitylene	4	99
cyclohexyl-CO$_2$H	Ph~NH$_2$	Xylene / Mesitylene	48[b] / 24[b]	91[b] / 98[b]
adamantyl-CO$_2$H	Ph~~NH$_2$	Mesitylene	2	92
Ph~=~CO$_2$H	HN (3,5-dimethylpiperidine)	Xylene	29	99
PhCO$_2$H	HN (3,5-dimethylpiperidine)	Mesitylene	20	95
Ph‑CH(OH)‑CO$_2$H	Ph~NH$_2$	Toluene	10	95 (94% ee)
i-Bu‑CH(OH)‑CO$_2$H	Ph~NH$_2$	Toluene	10	87 (>98% ee)
i-Pr‑CH(OH)‑CO$_2$H	Ph~NH$_2$	Toluene	10	96 (>98% ee)

[a]Unless otherwise noted, results taken from reference 5. [b]The reaction was carried out using amines (30 mmol) and carboxylic acids (33 mmol) in the presence of (3,4,5-trifluorophenyl)boric acid (0.3 mmol) in solvents (60 mL) by heating under refulx with removal of water.

TABLE II
LACTAMIZATION REACTION OF AMINOCARBOXYLIC ACIDS CATALYZED BY (3,4,5-TRIFLUOROPHENYL)BORONIC ACID[a]

AminocarBoxylic Acid	Solvent	Time (hr)	Product	Yield (%)
HO$_2$C, H-N pyrrolidine structure	Anisole	11	bicyclic diketopiperazine structure	94
HO$_2$C~~~NH$_2$	Xylene	18	piperidinone (NH) structure	20
HO$_2$C~~~~~NH$_2$	Xylene	22	azepanone (NH) structure	93

[a]Unless otherwise noted, results taken from reference 5.

185

PREPARATION OF SECONDARY AMINES FROM PRIMARY AMINES VIA 2-NITROBENZENESULFONAMIDES:

N-(4-METHOXYBENZYL)-3-PHENYLPROPYLAMINE

[Benzenepropanamine, N-[(4-methoxyphenyl)methyl]-]

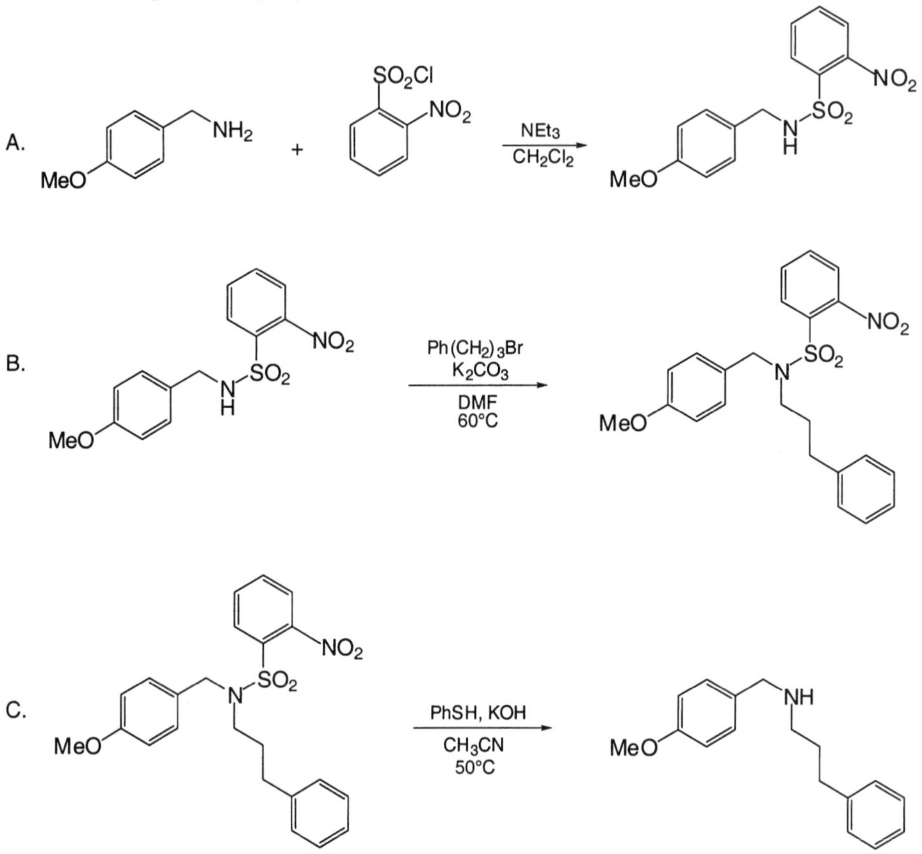

A.

B.

C.

Submitted by Wataru Kurosawa, Toshiyuki Kan, and Tohru Fukuyama.[1]

Checked by Audra M. Dalton and Rick L. Danheiser.

1. Procedure

A. N-(4-Methoxybenzyl)-2-nitrobenzenesulfonamide. A 300-mL, two-necked, round-bottomed flask equipped with a magnetic stirring bar, nitrogen gas inlet, and a rubber septum is charged with 6.81 g (49.6 mmol) of 4-methoxybenzylamine (Note 1), 100 mL of dichloromethane and 6.93 mL (49.6 mmol) of triethylamine (Note 1). The mixture is stirred and cooled in an ice-water bath while 10.0 g (45.1 mmol) of 2-nitrobenzenesulfonyl chloride (Note 1) is added over a period of 5 min. After 5 min, the ice bath is removed and the reaction mixture is allowed to warm to room temperature, stirred for 15 min (Note 2), and then quenched with 100 mL of 1N hydrochloric acid (HCl). The aqueous layer is extracted with two 100-mL portions of dichloromethane, and the combined organic extracts are washed with 50 mL of brine, dried over magnesium sulfate, filtered, and concentrated under reduced pressure to give 14.2 g (98%) of the crude 2-nitrobenzenesulfonamide. Recrystallization from 500 mL of 1:1 ethyl acetate/hexane gives 13.00-13.15 g (90-91%) of N-(4-methoxybenzyl)-2-nitrobenzenesulfonamide as white crystals (Note 3).

B. N-(4-Methoxybenzyl)-N-(3-phenylpropyl)-2-nitrobenzenesulfonamide. A 200-mL, two-necked, round-bottomed flask equipped with a magnetic stirring bar, a nitrogen gas inlet, and a rubber septum is charged with 10.0 g (31.0 mmol) of N-(4-methoxybenzyl)-2-nitrobenzenesulfonamide, 12.9 g (93.1 mmol) of potassium carbonate (Note 4), and 40 mL of anhydrous dimethylformamide (DMF). To the stirred mixture is added 5.19 mL (34.1 mmol) of 3-phenylpropyl bromide (Note 5) over a period of 5 min and the resulting mixture is heated in a 60°C oil bath for 70 min (Note 6). The reaction mixture is allowed to cool to room temperature, diluted with 250 mL of water, and extracted with three 250-mL portions of ether. The combined organic extracts are washed with brine (100 mL), dried over magnesium sulfate, filtered, and concentrated under reduced pressure to give a pale yellow liquid. The residue is purified by column chromatography on silica gel (Note 7) to give   13.5   g   (99%)   of   N-(4-methoxybenzyl)-N-(3-phenylpropyl)-2-nitrobenzenesulfonamide (Note 8) as a viscous pale yellow liquid.

C. N-(4-Methoxybenzyl)-3-phenylpropylamine. A 100-mL, two-necked, round-bottomed flask equipped with a magnetic stirring bar, nitrogen gas inlet, and a rubber septum is charged with 7.82 mL (76.5 mmol) of thiophenol (Note 9) and 20 mL of acetonitrile (CH_3CN). The mixture is cooled in an ice-water bath and 10.9 M aqueous potassium hydroxide solution (7.02 mL, 76.5 mmol) is added over a period of 10 min. After 5 min, the ice-water bath is removed, and 13.5 g (30.6 mmol) of N-(4-methoxybenzyl)-N-(3-phenylpropyl)-2-nitrobenzenesulfonamide in 20 mL of acetonitrile is added over 20 min. The reaction mixture is heated in a 50°C oil bath for 40 min (Note 10). The reaction mixture is allowed to cool to room temperature, diluted with 80 mL of water, and extracted with three 80-mL portions of dichloromethane. The combined organic extracts are washed with brine (80 mL), dried over magnesium sulfate, filtered, and concentrated under reduced pressure. The residue is purified by column chromatography on silica (Note 11) to give 7.81 g of the desired amine and its hydrochloride salt. This oil is dissolved in 120 mL of dichloromethane and washed with two 80-mL portions of 1 M aqueous sodium hydroxide solution, 40 mL of brine, dried over magnesium sulfate, filtered, and concentrated under reduced pressure. Bulb-to-bulb distillation (0.25 mm, oven temperature 150°C) provides 6.98-7.08 g (89-91%) of N-(4-methoxybenzyl)-3-phenylpropylamine as a colorless oil (Notes 12, 13).

2. Notes

1. 4-Methoxybenzylamine, triethylamine, and 2-nitrobenzenesulfonyl chloride were purchased by the submitters from Tokyo Kasei Kogyo Co. The checkers obtained 4-methoxybenzylamine and 2-nitrobenzenesulfonyl chloride from Alfa Aesar and triethylamine from Mallinckrodt Inc..

2. All reactions were monitored by TLC analysis on Merck silica gel 60 F254 plates, which were visualized by a 254-nm UV lamp and stained with an ethanolic solution of phosphomolybdic acid.   TLC analysis showed clean formation of the 2-nitrobenzenesulfonamide (hexane : ethyl acetate 3 : 2, R_f = 0.33).

3. Yield is based on 2-nitrobenzenesulfonyl chloride.   The crude product was practically pure as judged by ^{1}H NMR analysis and may be used for the next step without purification. The recrystallized compound exhibits the following properties: mp 123°C; ^{1}H NMR (400 MHz, CDCl$_3$) δ: 3.76 (s, 3 H), 4.25 (d, 2 H, J = 6.2), 5.63 (br, t, 1 H, J = 6.2), 6.75 (d, 2 H, J = 8.5), 7.13 (d, 2 H, J = 8.5), 7.63-8.03 (m, 4 H); ^{13}C NMR (100 MHz, CDCl$_3$) δ: 47.4, 55.3, 114.0, 125.2, 127.7, 129.2, 131.1, 132.7, 133.3, 134.0, 159.3; IR (thin film) cm^{-1}: 3312, 2941, 1543, 1511, 1363, 1337, 1243, 1160; MS m/z: 322, 134, 121. Anal. Calcd for C$_{14}$H$_{14}$N$_2$O$_5$S: C, 52.17; H, 4.38; N, 8.69. Found: C, 52.05; H, 4.46; N, 8.74.

4. The checkers obtained anhydrous DMF from EM Sciences.   Potassium carbonate (powder, K$_2$CO$_3$) was purchased from Aldrich Chemical Company, Inc.   If granular K$_2$CO$_3$ is used in place of powder, the reaction requires a longer time (5.5 hr) and proceeds in lower yield (81%).

5. 3-Phenylpropyl bromide was purchased from Tokyo Kasei Kogyo Co. or Alfa Aesar.

6. TLC analysis showed clean formation of the alkylated sulfonamide (hexane : ethyl acetate 1 : 1, R_f = 0.71).

7. Column chromatography was performed on 150 g of silica gel (100-210 μm, Kanto Chemical Co., Inc. or Silacycle, Inc.). The product was eluted with 300 mL of 10% ethyl acetate-hexane, 300 mL of 25% ethyl acetate-hexane, and 1.8 L of 40% ethyl acetate-hexane, and 300-mL fractions were collected.

8. The product exhibits the following properties: ^{1}H NMR (400 MHz, CDCl$_3$) δ: 1.70 (dt, 2 H, J = 7.7, 7.7), 2.44 (t, 2 H, J = 7.7), 3.23 (t, 2 H, J = 7.7), 3.79 (s, 3 H), 4.44

(s, 2 H), 6.81 (d, 2 H, J = 8.7), 6.99 (d, 2 H, J = 8.7), 7.14-7.25 (m, 5 H), 7.58-7.92 (m, 4 H); ^{13}C NMR (100 MHz, CDCl$_3$) δ: 29.0, 32.6, 46.4, 50.7, 55.2, 114.0, 124.1, 125.9, 127.5, 128.2, 128.3, 129.7, 130.7, 131.6, 133.3, 133.6, 140.9, 147.8, 159.5; IR (neat) cm^{-1}: 2934, 1543, 1513, 1372, 1346, 1250, 1211; MS m/z 440, 150, 122. Anal. Calcd for C$_{23}$H$_{24}$N$_2$O$_5$S: C, 62.71; H, 5.49; N, 6.36. Found: C, 62.76; H, 5.47; N, 6.31.

9. Thiophenol and potassium hydroxide were purchased by the submitters from Tokyo Kasei Kogyo Co. and by the checkers from Aldrich Chemical Company, Inc. and Mallinckrodt Inc., respectively.

10. TLC analysis showed clean formation of the deprotected amine (methanol:dichloromethane 10 : 90, R$_f$ = 0.52).

11. Column chromatography was performed on 150 g of silica gel (100-210 μm, Kanto Chemical Co., Inc. or Silacycle, Inc.). The product was eluted with 70 mL of dichloromethane, 900 mL of 2% methanol-dichloromethane, and 1.8 L of 2.5:2.5:95 isopropylamine-methanol-dichloromethane (300-mL fractions).

12. The product exhibits the following properties: ^{1}H NMR (400 MHz, CDCl$_3$) δ: 1.83 (dt, 2 H, J = 7.8, 7.8), 2.65 (m, 4 H), 3.71 (s, 2 H), 3.79 (s, 3 H), 6.84-7.28 (m, 9 H); ^{13}C NMR (100 MHz, CDCl$_3$) δ: 31.6, 33.6, 48.7, 53.4, 55.1, 113.8, 125.8, 128.3, 129.2, 132.6, 142.1, 158.5; IR (neat) cm^{-1}: 3302, 2931, 1511, 1246; MS m/z 255, 150, 121. Anal. Calcd for C$_{17}$H$_{21}$NO: C, 79.96; H, 8.29; N, 5.49. Found: C, 79.76; H, 8.40; N, 5.41.

13. The amine can be transformed to the hydrochloride salt by bubbling a stream of hydrogen chloride gas into a solution of 7.81 g of the amine in methanol at 0°C. Recrystallization from 2-propanol gives N-(4-methoxybenzyl)-3-phenylpropylamine hydrochloride (7.92 g, 88%) as white crystals. The product exhibits the following properties: mp 206°C; ^{1}H NMR (400 MHz, CDCl$_3$) δ: 2.14 (dt, 2 H, J = 7.4, 7.7), 2.62 (t, 2 H, J = 7.4), 2.73 (t, 2 H, J = 7.7), 3.74 (s, 3 H), 3.91 (s, 2 H), 6.87 (d, 2 H, J = 8.6), 7.10-7.24 (m, 5 H), 7.45 (d, 2 H, J = 8.6), 9.80 (br, s, 1 H); ^{13}C NMR (100 MHz, CDCl$_3$) δ: 27.0, 32.5, 44.9, 49.8, 55.1, 114.2, 121.8, 126.1, 128.2, 128.4, 131.8, 139.7, 160.2;

IR (thin film) cm^{-1}: 2938, 2789, 1518, 1252; MS m/z 255, 150, 121. Anal. Calcd for C$_{17}$H$_{22}$ClNO: C, 69.97; H, 7.60; N, 4.80. Found: C, 69.85; H, 7.58; N, 4.86.

Waste Disposal Information

All toxic materials were disposed of in accordance with "Prudent Practices in the Laboratory"; National Academy Press; Washington, DC, 1995.

3. Discussion

Conversion of primary amines to the corresponding secondary amines appears to be deceptively simple.[2] Alkylation of primary amines with alkyl halides or sulfonates frequently leads to the formation of the undesired tertiary amines and/or quaternary ammonium salts. Reductive alkylation with aldehydes or ketones using sodium cyanoborohydride (NaBH$_3$CN) often produces tertiary amines to a varying extent unless the desired secondary amine is sterically hindered. Reduction of N-monoalkyl amides with such strong reducing agents as lithium aluminum hydride (LiAlH$_4$), diisobutylaluminum hydride (DIBAL), or borane seems to be the most reliable procedure. To circumvent these problems, the Mitsunobu alkylations of toluenesulfonamides[3] and trifluoroacetamides[4] have recently been reported. However, because of the relatively harsh deprotection conditions, these methods do not appear to be suitable for the preparation of the base-sensitive secondary amines. The present procedure describes the simple and general transformation of primary amines to the corresponding secondary amines using the 2-nitrobenzenesulfonamide protecting group that can be applied to the synthesis of a wide range of secondary amines (Scheme 1).[5] A related procedure using 2,4-dinitrobenzenesulfonamides that requires even milder deprotection conditions (HSCH$_2$CO$_2$H, Et$_3$N, CH$_2$Cl$_2$, room temperature) has recently been reported.[6]

191

Scheme 1

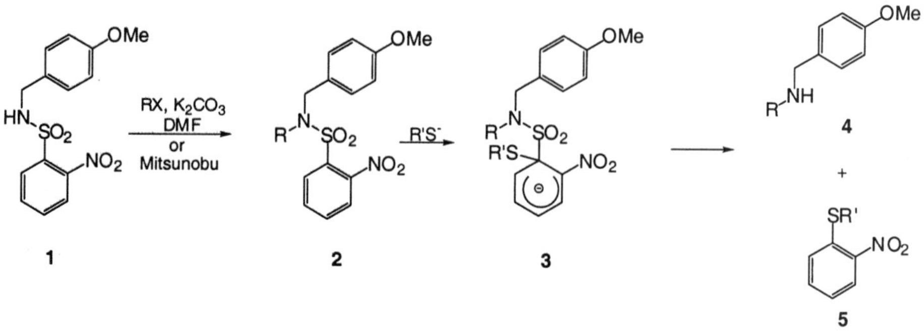

Protection of the primary amines was performed by treatment with 2-nitrobenzenesulfonyl chloride and base (triethylamine, pyridine, or 2,6-lutidine) to give N-monosubstituted 2-nitrobenzenesulfonamides in high yields (Step A). Alkylation of N-monosubstituted 2-nitrobenzenesulfonamides (1) proceeded smoothly under either the conditions described above (conventional) or Mitsunobu conditions[7] to give N,N-disubstituted 2-nitrobenzenesulfonamide (2) in excellent yields. For large-scale alkylations, conventional conditions are recommended, because of the ease of purification. Facile deprotection of N,N-disubstituted 2-nitrobenzenesulfonamides is achieved by treatment with thiolate nucleophile, presumably via the formation of a Meisenheimer complex[8] (3), giving the desired secondary amines (4) in excellent yields (Step C). Since potassium hydroxide is inexpensive, the described procedure is convenient for a large-scale reaction. For a small scale reaction, however, one of the following reported procedures is recommended: (1) potassium carbonate, thiophenol in DMF, (2) cesium carbonate, thiophenol in CH_3CN, (3) lithium hydroxide, mercaptoacetic acid in DMF. Procedure (3) has the advantage that the by-product 2-nitrophenylthioacetic acid (5) can be easily removed by partitioning between ether and an aqueous sodium bicarbonate solution. Representative examples of this protocol are summarized in Table I.

Since the 2-nitrobenzenesulfonamide group is stable under acidic [HCl (10 eq), MeOH, 60°C, 4 hr] as well as basic [NaOH (10 eq), MeOH, 60°C, 4 hr] conditions, it can be used extensively for protection of primary and secondary amines. Because of the mild conditions and easy procedure, the submitters believe that the use of 2-nitrobenzenesulfonamides serves as a method of choice for the preparation of a wide variety of secondary amines comparable to the Gabriel synthesis for primary amines.

1. Graduate School of Pharmaceutical Sciences, The University of Tokyo, 7-3-1. Hongo, Bunkyo-ku, Tokyo 113-0033, Japan.

2. For general syntheses of amines, see: Sandler, S. R.; Karo, W. "Organic Functional Group Preparations", 2nd ed.; Academic, New York, 1983; Chapter 13.

3. Henry, J. R.; Marcin, L. R.; McIntosh, M. C.; Scola, P. M.; Harris, G. D. Jr.; Weinreb, S. M. *Tetrahedron Lett.* **1989**, *30*, 5709.

4. Tsunoda, T.; Otsuka, J.; Yamamiya, Y.; Ito, S. *Chem. Lett.* **1994**, 539.

5. Fukuyama, T.; Jow, C.-K.; Cheung, M. *Tetrahedron Lett.* **1995**, *36*, 6373.

6. Fukuyama, T.; Cheung, M.; Jow, C.-K.; Hidai Y.; Kan, T. *Tetrahedron Lett.* **1997**, *38*, 5831.

7. (a) Mitsunobu, O. *Synthesis* **1981**, 1; (b) Hughes, D. L. *Org. React.* **1992**, *42*, 335.

8. (a) Terrier, F. *Chem. Rev.* **1982**, *82*, 77; (b) Artamkina, G. A.; Egorov, M. P.; Beletskaya, I. P. *Chem. Rev.* **1982**, *82*, 427.

Appendix

Chemical Abstracts Nomenclature (Collective Index Number);

(Registry Number)

N-(4-Methoxybenzyl)-3-phenylpropylamine: Benzenepropanamine,

N-[(4-methoxyphenyl)methyl]- (13); (145060-50-0)

N-(4-methoxybenzyl)-2-nitrobenzenesulfonamide: Benzenesulfonamide, N-[(4-

methoxyphenyl)methyl]-2-nitro- (13); (171414-16-7)

4-Methoxybenzylamine: Benzylamine, p-methoxy- (8); Benzenemethaneamine,

 4-methoxy- (9); (2393-23-9)

Triethylamine (8): Ethanamine, N,N-diethyl- (9); (121-44-8)

o-Nitrobenzenesulfonyl chloride: Benzenesulfonyl chloride, o-nitro- (8); Benzenesulfonyl

chloride, 2-nitro- (9); (1694-92-4)

3-Phenylpropyl bromide: ALDRICH: 1-Bromo-3-phenylpropane: Benzene,

(3-bromopropyl)- (8,9); (637-59-2))

Thiophenol: Benzenethiol (13); (108-98-5)

TABLE 1

ALKYLATION AND DEPROTECTION OF 2-NITROBENZENESULFONAMIDES

RX or ROH	Alkylation conditions[a]	2[b] (% isolated yield)	Deprotection conditions[c]	4[b,d] (%isolated yield)
Ph⌒Br	A	Ph⌒N(PMB)(SO$_2$Ar) (98)	TP	Ph⌒N(H)(PMB) (94)
			MA	(93)
⌒⌒Br	B	⌒⌒N(PMB)(SO$_2$Ar) (98)	TP	⌒⌒N(H)(PMB) (94)
Ph⌒⌒OH	C	Ph⌒⌒N(PMB)(SO$_2$Ar) (91)	TP	Ph⌒⌒N(H)(PMB) (88)
(CO$_2$Et)(OH)	C	EtO$_2$C⌒N(PMB)(SO$_2$Ar) (87)	TP	EtO$_2$C⌒N(H)(PMB) (93)

[a]A: RX (1.1 eq), K$_2$CO$_3$ (2 eq), DMF, 23°C, 1 hr.
B: RX (1.1 eq), K$_2$CO$_3$ (2 eq), DMF, 60°C, 30 min.
C: ROH (1.3 eq), DEAD (1.3 eq), PPh$_3$ (1.3 eq), CH$_2$Cl$_2$, 23°C, 1 hr.
[b]Satisfactory spectroscopic data were obtained on all new compounds.
[c]TP: PhSH (1.2 eq), K$_2$CO$_3$ (3 eq), DMF, 23°C, 40 min.
MA: HSCH$_2$CO$_2$H (2 eq), LiOH (4 eq), DMF, 23 °C, 1 hr.
[d]Separated by silica gel chromatography after partitioning between Et$_2$O and a dilute NaHCO$_3$ solution.

2-AMINO-3-FLUOROBENZOIC ACID

[Benzoic acid, 2-amino-3-fluoro-]

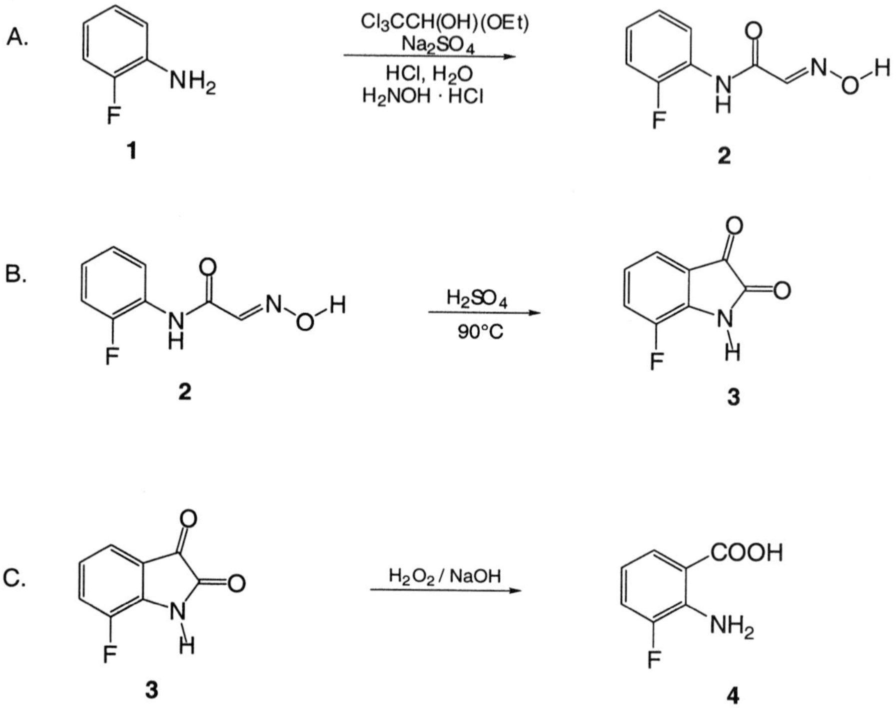

A.

$Cl_3CCH(OH)(OEt)$
Na_2SO_4
HCl, H_2O
$H_2NOH \cdot HCl$

1 → **2**

B.

H_2SO_4
90°C

2 → **3**

C.

$H_2O_2 / NaOH$

3 → **4**

Submitted by Martin Kollmar, Richard Parlitz, Stephan R. Oevers, and Günter Helmchen.[1]

Checked by Hui Li and Marvin J. Miller.

1. Procedure

A. N-(2-Fluorophenyl)-2-(hydroxyimino)acetamide (**2**). Solution A: A 2-L, three-necked, round-bottomed flask fitted with a condenser and a thermometer is charged with 62.0 g (0.89 mol) of hydroxylamine hydrochloride, 256.7 g (1.80 mol) of anhydrous

196

sodium sulfate, 79.5 g (0.41 mol) of 2,2,2-trichloro-1-ethoxyethanol (Note 1) and 1125 mL of water. To aid dissolution, the mixture is heated to approximately 40°C and stirred vigorously with the help of a mechanical stirrer (Note 2). Solution B: 30 g (0.27 mol) of 2-fluoroaminobenzene (Note 3) is added dropwise slowly into a 500-mL, one-necked, round-bottomed flask containing a vigorously stirred mixture of 150 mL of water and 75 mL of concd hydrochloric acid (Note 4). Solution B is added in one portion to solution A. The mixture is vigorously stirred and heated to reflux. After 1 to 2 min the mixture turns milky and a white precipitate accompanied by a small amount of brown by-product is formed (Note 5). The oil bath is removed and the flask is cooled rapidly (ice bath) to room temperature (20°C) (Note 6). After 60 hr at room temperature the precipitate is removed by filtration and washed with ice-cold water (Note 7). After drying over phosphorus pentoxide (P_4O_{10}), 43.6 g (86%) of product, mp 116-117°C, is obtained (Note 8). Crystals glued together by brown by-product, mainly consisting of the product, can be used in the next step without further purification. Nearly colorless product is obtained by recrystallization from ethanol.

B. *7-Fluoroisatin* (**3**). A 250-mL, three-necked, round-bottomed flask fitted with a condenser and a thermometer is charged with 100 mL of concd sulfuric acid. After heating to 70°C, 30.0 g (0.165 mol) of anilide **2** (Note 9) is added over a period of 1 hr. The resulting deep red solution is heated to 90°C (Note 10) for 60 min (Note 11) and then is cooled to room temperature (20°C) over an ice bath (Note 12). The mixture is then added rapidly to a vigorously stirred mixture of 1.0 L of ice water and 200 mL of ethyl acetate (Note 13). The organic phase is separated and the almost black aqueous phase is extracted twice with 200 mL of ethyl acetate (Note 14). The combined red organic phases are dried with sodium sulfate. The solvent is removed under reduced pressure and the crude product is dried at low pressure, whereupon 12.9 to 15.7 g (47-57%) of an orange powder, mp 186-190°C, is obtained (Note 15). The crude product is sufficiently pure for the next step. Further purification is possible by recrystallization from acetone/water.

197

C. 2-Amino-3-fluorobenzoic acid (**4**). A 500-mL, three-necked, round-bottomed flask fitted with an addition funnel and a thermometer, is charged with 15.0 g (0.09 mol) of 7-fluoroisatin (**3**) and 200 mL of 1 M aqueous sodium hydroxide solution (Note 16); 22 mL of hydrogen peroxide (30%) solution (0.20 mol hydrogen peroxide) is added dropwise over 45 min. The temperature of the reaction mixture rises to 30° or 40°C. After 1.5 hr the reaction is complete (Note 17). To the pale orange, clear reaction mixture 3 M hydrochloric acid is added until a pH of ca. 7.5 is reached. The mixture is treated with charcoal, stirred for a while, filtered and the clear filtrate is further acidified to pH 4-5, when the now pale yellow solution becomes cloudy again. Finally, at pH 1 the beige 3-fluoroanthranilic acid (**4**) precipitates. Bubbles are observed during acidification. After an hour of stirring, the product is collected on a funnel and dried over P_4O_{10}; yield: 11.64 to 13.3 g (84-96%) of pure 3-fluoroanthranilic acid (**4**), mp 182-184°C (Note 18).

2. Notes

1. *CAUTION:* *2,2,2-Trichloro-1-ethoxyethanol is toxic.* 2,2,2-Trichloro-1-ethoxyethanol was obtained from Aldrich Chemical Company, Inc.

2. The salts can also be dissolved at room temperature. Warming to ca. 40°C facilitates dissolution. Note that complete dissolution and observance of precise concentration of the solution are essential for the success of the procedure.

3. The quality of the commercially available (colorless) 2-fluoroaminobenzene (**1**) is sufficient (Fluka Chemical Company). Older or colored material requires distillation prior to use (bp 171°C, d = 1.15).

4. Dissolution of aniline **1** is exothermic and it should therefore be added in small portions. Complete dissolution is essential in order to avoid the formation of dark tar-like by-products.

198

5. The longer the solution boils the more tar-like by-product is formed. One to two min of boiling are necessary and sufficient for complete conversion of the reactants.

6. The reaction mixture must not be cooled to 0°C as this leads to the precipitation of inorganic salts.

7. The long period is necessary to obtain maximum yield.

8. Spectral characteristics are as follows: IR (KBr) cm^{-1} : 3390 m (O-H), 1660 s (C=O), 1618 s (C=N), 1546, 1486, 1460 s (C=C), 1260 s (C-F)], 1021 m [ν(N-O)], 756 s [ν(C-H)$_{arom}$]; ^{1}H NMR (CD$_3$OD, 500.13 MHz) δ: 7.09 (m, 3 H, H-3, H-4, H-5), 7.53 (s, 1 H, H$\underline{C}$=NOH), 7.94 (m, 1 H, H-6); ^{13}C NMR (CD$_3$OD,75.47 MHz) δ: 116.33 (d, $^2J_{F,3}$ = 19.8, C-3), 124.76 (C-5), 125.47 (d, $^3J_{F,6}$ = 4.0, C-6), 126.53 (d, $^2J_{F,1}$ = 11.3, C-1), 126.99 (d, $^3J_{F,4}$ = 7.9, C-4), 144.01 (H$\underline{C}$=NOH), 155.42 (d, $^1J_{F,2}$ = 245.3, C-2), 162.88 ($\underline{C}$=O).

9. Anilide 2 has to be completely dry. Residual water reacts violently with the acid with heat generation causing decomposition.

10. By-products form if the temperature is too high. Anilide 2 does not dissolve completely if the temperature is below 50°C and then the reaction does not go to completion.

11. The progress of the reaction can be monitored by hydrolysis of a sample, extraction with ethyl acetate, and TLC [silica gel Macherey, Nagel & Co. "Polygram Sil G/UV 254", petroleum ether/ethyl acetate/acetic acid 99:50:1, UV visualization, R$_f$ (2) = 0.40, R$_f$ (3) = 0.31 (yellow spot)].

12. If the temperature is too high, tar-like by-products form. If the solution is cooled to 0°C, hydrolysis does not take place because the sulfuric acid does not mix with the hydrolysis solution and mainly oxime 5 is obtained.

13. The presence of ethyl acetate is essential as otherwise the yellow oxime 5 (mp 233-235°C) is formed in yields of 20-30%. Ethyl acetate is added in order to extract the isatin 3 from the aqueous phase immediately upon formation. Oxime 5 is probably formed

by reaction of isatin **3** with hydroxylamine generated by decomposition of unreacted anilide **2**.[2]

5

Spectral characteristics of 7-fluoroisatin 3-oxime (**5**) are as follows: TLC: silica gel Macherey, Nagel & Co. "Polygram Sil G/UV 254", petroleum ether/ethyl acetate PE/EE 2:1 elution, 2 drops of glacial acetic acid, UV visualization, $R_f = 0.19$; IR (KBr) cm^{-1}: ~3500 m, br (OH), 1723 s (C=O), 1640 s (C=N), 1596, 1494, 1445 m (C=C), 1208 m (C-F), 942 m (N-O), 794 w (C-H$_{arom}$); ^{1}H NMR (acetone-d$_6$, 500.13 MHz) δ: 7.06 (ddd, 1 H, $^4J_{5,F} + 4.6$, $^3J_{4,5} = 7.5$, $^3J_{5,6} = 8.5$, H-5), 7.20 (ddd, 1 H, $^4J_{4,6} = 1.1$, $^3J_{5,6} = 8.5$, $^3J_{6,F} = 10.1$, H-6), 7,83 (dd, 1 H, $^4J_{4,6} = 1,0$, $^3J_{4,5} = 7.0$, H-4), 10.14 (bs, 1 H, N-H̲), 12.75 (bs, 1H, N-OH̲); ^{13}C NMR (acetone-d$_6$, 125.76 MHz) δ: 115.75 (d, $^3J_{F,3} = 4.2$, C-3a), 115.96 (C-6), 120.33 (C-5), 120.43 (C-4), 126.40 (d, $^2J_{F,8} = 13.4$, C-7a), 140.65 (d, $^4J_{F,2} = 4.1$, C-3), 143.85 (d, $^1J_{F,7} = 243,2$, C-7), 161.71 (C-2); HR-MS (EI, direct insert): m/z 180.03322 (M$^+$ exact mass calcd for C$_8$H$_5$O$_2$N$_2$F: 180.0335), 163.0070 (M$^+$ - OH), 152.0365 (M$^+$ - CO), 135.0359 (M$^+$ - CO$_2$H), 108.0271 (M$^+$ - CO – OH – HCN). Anal. Calcd for C$_8$H$_5$FN$_2$O$_2$: C, 53.43; H, 2.80; N, 15.55. Found: C, 53.34; H, 3.08; N, 15.28.

14.   If the aqueous phase is extracted more than twice, the yield may rise, but the oxime **5** and other by-products are extracted as well.

15.   Spectral characteristics are as follows: IR (KBr) cm^{-1}: 3446 w (NH̲-CO), 1737 s (C3=O), 1637 s (C2=O), 1602, 1495, 1452 m (C=C), 1209 m (C-F), 780 w (C-H)$_{arom}$; ^{1}H NMR (acetone-d$_6$, 500.13 MHz) δ: 7.14 (ddd, 1 H, $^4J_{5,F} = 4.2$, $^3J_{4,5} = 7.7$, $^3J_{5,6} = 8.3$, H-5), 7.39 (dd, 1 H, $^4J_{4,6} = 1.1$, $^3J_{4,5} = 7.4$, H-4), 7.48 (ddd, 1 H, $^4J_{4,6} = 1.0$, $^3J_{4,5} = 8.4$, $^3J_{6,F} = 10.2$, H-6), 10.45 (bs, 1 H, N-H); ^{13}C NMR (acetone-d$_6$, 125.76 MHz) δ: 119.15 (d, $^3J_{F,3} = 3.4$, C-3a), 119.26 (d, $^4J_{F,4} = 3.5$, C-4), 122.52 (d, $^3J_{F,5} = 5.3$, C-5),

123.57 (d, $^2J_{F,6}$ = 17.6, C-6), 136.20 (d, $^2J_{F,8}$ = 13.9, C-7a), 146.38 (d, $^1J_{F,7}$ = 246.8, C-7), 157.29 (C-2), 181.70 (d, $^4J_{F,2}$ = 4.5, C-3).

16. Previously[3] 3 N or 10 N NaOH was used. The submitters found that this reaction also proceeds to completion at room temperature in 1 N NaOH solution.

17. Monitoring was carried out by extracting acidified samples with ethyl acetate and TLC [silica gel Macherey, Nagel & Co. "Polygram Sil G/UV 254", petroleum ether/ethyl acetate/acetic acid 99:50:1, visualization of spots by UV R_f = 0.37 (spot shows strong blue fluorescence)].

18. Spectral characteristics of **4** are as follows: IR (KBr) cm^{-1}: 3500-3391 m (NH$_2$), 3085 m, br (CO$_2$-H), 1678 s (C=O), 1630 w (C-NH$_2$)], 1590, 1562, 1476 m (C=C)], 780 w (C-H$_{arom}$); ^{1}H NMR (acetone-d$_6$, 500.13 MHz) δ: 6.58 (ddd, 1 H, $^4J_{5,F}$ = 4.9, $^3J_{5,6}$ = 7.9, $^3J_{4,5}$ = 8.0, H-5), 6.84 (bs, 1 H), 7.19 (ddd, 1 H, $^4J_{4,6}$ = 1.3, $^3J_{4,5}$ = 8.0, $^4J_{4,F}$ = 11.6, H-4), 7.31 (bdd, 1 H), 7.63 (bd, 1 H), 7.67 (ddd, 1 H, $^3J_{6,F}$ = 1.3, $^4J_{4,6}$ = 1.3, $^3J_{5,6}$ = 7.9, H-6); ^{13}C NMR (acetone-d$_6$, 125.76 MHz) δ: 108.71 (d, $^3J_{F,1}$ = 4.5, C-1), 110.30 (d, $^3J_{F,5}$ = 7.0, C-5), 114.51 (d, $^2J_{F,4}$ = 18.5, C-4), 122.97 (d, $^4J_{F,6}$ = 3.0, C-6), 136.35 (d, $^2J_{F,2}$ = 14.0, C-2), 147.55 (d, $^1J_{F,3}$ = 237.4, C-3), 165.27 (d, $^4J_{F,7}$ = 3.4, C-7).

Waste Disposal Information

All toxic materials were disposed of in accordance with "Prudent Practices in the Laboratory"; National Academy Press; Washington, DC, 1995.

3. Discussion

Anthranilic acids are important intermediates for the preparation of heterocycles. The previously described procedure,[3] used here as the starting point, works well for compounds not containing electron-withdrawing substituents. This modified procedure thus extends the range of applicability.

2-Amino-3-fluorobenzoic acid is an important intermediate in the synthesis of derivatives of indole, such as the potent and selective thromboxane/prostaglandin endoperoxide receptor antagonist L-670,596[4] or the anti-inflammatory agent Etodolac.[5] Compounds of this type have therapeutic applications. 2-Amino-3-fluorobenzoic acid is also an important precursor for the synthesis of fluoroacridines, which can be converted to interesting tridentate ligands, such as Acriphos.[6]

| L-670,596 | Etodolac | Acriphos |

The steps described above at least triple yields that were previously reported;[3] in particular the yield of the second step is improved significantly. No chromatography is required for purification and all reactions can be carried out on a larger scale, the only limiting factor being the scale of the laboratory equipment. Of advantage is the use of water as solvent in all three steps.

In order to assess the generality of this procedure for the preparation of acceptor-substituted anthranilic acids it was applied to 2-amino-3-chlorobenzoic acid, which was obtained with excellent overall yield of 53% (lit.[3]: 16%).

1. Organisch-Chemisches Institut der Universität Heidelberg, Im Neuenheimer Feld 270, D-69120 Heidelberg, Germany and (S.R.O.) Max-Planck-Institut für Kohlenforschung, Kaiser-Wilhelm-Platz 1, D-45470 Mühlheim an der Ruhr, Germany.

2. Wibaut, J. P.; Geerling, M. C. *Rec. Trav. Chim.* **1931**, *50*, 41.

3. (a) Holt, S. J.; Sadler, P. W. *Proc. Roy. Soc. B* **1958**, *148*, 481; (b) McKittrick, B.; Failli, A.; Steffan, R. J.; Soll, R. M.; Hughes, P.; Schmid, J.; Asselin, A. A.; Shaw, C.C.; Noureldin, R.; Gavin, G. *J. Heterocycl. Chem.* **1990**, *27*, 2151.

4. Ford-Hutchinson, A. W.; Girard, Y.; Lord, A.; Jones, T. R.; Cirino, M.; Evans, J. F.; Gillard, J.; Hamel, P.; Leveille, C.; Masson, P.; Young, R. *Can. J. Physiol. Pharmacol,* **1989**, *67*, 989.

5. (a) Demerson, C. A.; Humber, L. G.; Philipp, A. H.; Martel, R. R. *J. Med. Chem.* **1976**, *19*, 391; (b) Humber, L. G.; *Med. Res. Rev.* **1987**, *7*, 1.

6. Hillebrand, S.; Bartkowska, B.; Bruckmann, J.; Krüger, C.; Haenel, M. W. *Tetrahedron Lett.* **1998**, *39*, 813.

Appendix
Chemical Abstracts Nomenclature (Collective Index Number);
(Registry Number)

2-Amino-3-fluorobenzoic acid: Anthranilic acid, 3-fluoro- (8); Benzoic acid, 2-amino-3-fluoro- (10); (825-22-9)

N-(2-Fluorophenyl)-2-(hydroxyimino)acetamide: Acetamide, N-(2-fluorophenyl)-2-(hydroxyimino)- (9); (349-24-6)

Hydroxylamine hydrochloride (8); Hydroxylamine, hydrochloride (9); (5470-11-1)

2,2,2-Trichloro-1-ethoxyethanol: Ethanol, 2,2,2-trichloro-1-ethoxy- (8,9); (515-83-3)

2-Fluoroaminobenzene: Aniline, o-fluoro- (8); Benzenamine, 2-fluoro- (9); (348-54-9)

7-Fluoroisatin: 1H-Indole-2,3-dione, 7-fluoro- (8,9); (317-20-4)

Hydrogen peroxide (8,9); (7722-84-1)

7-Fluoroisatin 3-oxime: 1H-Indole-2,3-dione, 7-fluoro-, 3-oxime (13); (143884-84-8)

3-(4-BROMOBENZOYL)PROPANOIC ACID
(Bromobutanoic acid, 4-bromo-γ-oxo-)

Submitted by Alexander J. Seed, Vaishali Sonpatki, and Mark R. Herbert.[1]

Checked by Ayako Ono and Koichi Narasaka.

1. Procedure

A 500-mL, three-necked, round-bottomed flask (Note 1) equipped with an overhead mechanical stirrer, is charged with powdered succinic anhydride (10.01 g, 0.1000 mol) (Note 2) and bromobenzene (96.87 g, 0.6170 mol) (Note 2) under dry argon. The resulting white mixture is cooled to 0°C before anhydrous aluminum chloride (26.67 g, 0.2000 mol) (Note 2) is added in one portion (Note 3). The reaction conditions are maintained over a period of 4 hr before the reaction mixture is allowed to warm to room temperature. The reaction mixture is stirred for 96 hr at room temperature (completion of the reaction is indicated by cessation of the evolution of hydrogen chloride gas) and is then poured into cooled (0°C), mechanically stirred hydrochloric acid (250 mL, 37%) (Note 4) and stirred for 1 hr. The white precipitate is filtered off, washed well with water (1 L) and dried overnight on a Büchner funnel. The crude product (24.81 g, 97%) is crystallized from dry toluene (Note 5) and dried under reduced pressure (P_2O_5, $CaCl_2$, 18 hr) to afford a white crystalline product (first fraction, 20.76 g, second fraction, 3.47 g); yield is 24.23 g (94%) (Note 6).

2. Notes

1. The glassware was dried in an oven at 130°C, assembled while still hot, and alternately evacuated and flushed with argon.

2. The checkers purchased succinic anhydride, bromobenzene and aluminum chloride from Wako Pure Chemical Industries, Ltd, Tokyo Chemical Industry Co., and Kanto Chemical Co. respectively, and used them as received.

3. Upon the addition of the aluminum chloride the reaction progressively turned from a yellow suspension to a clear yellow to a clear orange-red solution.

4. The checkers used 35-37% hydrochloric acid (340 mL) purchased from Kokusan Chemial Works, Ltd.

5. The toluene was dried over sodium metal.

6. The product showed the following physical and spectroscopic characteristics: mp 147-148°C (lit.[2] 149.5-150.2°C). ^{1}H NMR (500 MHz, DMSO-d_6) δ: 2.59 (2 H, t, ^3J = 6.5), 3.21 (2 H, t, ^3J = 6.5), 7.88 (2 H, d, ^3J = 8.8), 7.96 (2 H, d, ^3J = 8.8), 12.19 (1 H, s); ^{13}C NMR (126 MHz, DMSO-d_6) δ: 28.3, 33.6, 127.7, 130.3, 132.2, 135.9, 174.1, 198.2; IR (KBr) cm^{-1}: 3400-2600, 1730, 1670, 1585, 1479, 1447, 1410, 1332, 1281, 1241, 1198, 1105, 1074, 990, 905, 840, 791. MS m/z 256.2(M+), 185.1, 183.1(100%), 157.1, 155.1, 76.1, 75.1. Anal. Calcd for $C_{10}H_9BrO_3$: C, 46.72; H, 3.53. Found: C, 46.69; H, 3.49.

Waste Disposal Information

All toxic materials were disposed of in accordance with "Prudent Practices in the Laboratory"; National Academy Press; Washington, DC, 1995

3. Discussion

Recently we have reported the first highly efficient synthesis of ferroelectric liquid crystals bearing the 2-alkoxythiophene unit via Lawesson's reagent-mediated cyclization of γ-keto esters.[3,4] The requisite γ-keto acids were initially targeted through acylation of appropriately substituted aryl compounds using the procedure described by Fieser et al.[5] A literature search reveals that acylation of aryl units with succinic anhydride is almost always carried out using this procedure which requires elevated temperatures. Recent studies have shown that the acylation of bromophenyl systems using this procedure results in problematic debromination[2] and considerably lower yields.[6] In repeating the method of Fieser it was found that acylation of bromobenzene gave lower yields and a substantial quantity (8% by ^{1}H NMR) of the ortho product as well as the desired para isomer.

Our improved methodology uses low temperatures and extended reaction times, and has been shown to give consistently high yields of purified materials combined with complete regioselectivity in a variety of aryl systems (see Table I). The procedure is the first reliable, simple and general method for acylation of aryl systems using succinic anhydride. In addition, it was noted that the reaction conditions also gave the highest reported yield (previous yields range from 55-63%[7]) for acylation of bromothiophene with succinic anhydride. Of particular interest was the fact that we did not observe any ring opening that is so often reported in reactions involving combinations of aluminum chloride and thiophene derivatives.[8]

The methodology may also be applicable to the synthesis of pharmacodynamic agents based on the alkoxythiophene core.[9] Such materials should be readily accessible via our new cyclization methodology[3,4,10] which utilizes the γ-keto acid precursors.

1.   Department of Chemistry, W. H., Kent State University, Kent, OH 44242-0001.

2. Mallory, F. B.; Luzic, E.D.; Mallory, C. W; Carroll, P. J. *J. Org. Chem.* **1992**, *57*, 366.

3. Herbert, M. R.; Sonpatki, V. M.; Jákli, A; Seed, A. J. at *18th International Liquid Crystal Conference* (Sendai (Japan), July 2000).

4. Sonpatki, V.; Herbert, M. R.; Sandvoss, L.; Seed, A. J. at *220th ACS National Meeting* (Washington DC (USA), August 2000).

5. Fieser, L. F.; Seligman, A. M. *J. Am. Chem. Soc.* **1938**, *60*, 170.

6. Mathur, N. C.; Snow, M. S.; Young, K. M; Pincock, J. A. *Tetrahedron* **1985**, *41*, 1509.

7. Badger, G. M.; Rodda, H. J.; Sasse, W. H. F. *J. Chem. Soc.* **1954**, 4162.

8. Taylor, R. "Heterocyclic Compounds. Thiophene and its derivatives"; Gronowitz, S., Ed.; John Wiley & Sons: New York, **1986**, *44 (Part 4)*, 1-117.

9. Press, J. B. "Heterocyclic Chemistry. Thiophene and its derivatives"; Gronowitz, S., Ed.; John Wiley & Sons: New York, **1990**, *44 (Part 1)*, 397-456.

10. Herbert, M. R.; Sonpatki, V. M.; Jákli, A; Seed, A. J. *Mol. Cryst. Liq. Cryst.* In press.

Appendix

Chemical Abstracts Nomenclature (Collective Index Number); (Registry Number)

3-(4-Bromobenzoyl)propanoic acid: Benzenebutanoic acid, 4-bromo-γ-oxo- (9); (6340-79-0).

Succinic anhydride (8): 2,5-Furandione, dihydro- (9); (108-30-5).

Bromobenzene: Benzene, bromo- (8,9); (108-86-1).

Aluminum chloride (8,9); (7446-70-0).

TABLE I

FRIEDEL-CRAFTS ACYLATION OF AROMATIC COMPOUNDS USING SUCCINIC ANHYDRIDE

Product	Reaction Time	Purified Yield (%)
Br—C$_6$H$_4$—C(O)—CH$_2$CH$_2$—C(O)—OH	96 hr	94
CH$_3$—C$_6$H$_4$—C(O)—CH$_2$CH$_2$—C(O)—OH	216 hr	85
CH$_3$O—C$_6$H$_4$—C(O)—CH$_2$CH$_2$—C(O)—OH	120 hr	98
Br—(thiophene)—C(O)—CH$_2$CH$_2$—C(O)—OH	5 hr[a]	95

[a]Nitrobenzene was used as solvent (20 mL per 1.0 g of bromothiophene) and the reaction mixture was held at 0°C to -5°C for the entire period.

NUCLEOPHILIC AROMATIC SUBSTITUTION OF ARYL FLUORIDES BY SECONDARY NITRILES: PREPARATION OF 2-(2-METHOXYPHENYL)-2-METHYLPROPIONITRILE

[Benzeneacetonitrile, 2-methoxy-α,α-dimethyl-]

Submitted by Stéphane Caron,[1] Jill M. Wojcik, and Enrique Vazquez.

Checked by Li Dong and Marvin J. Miller

1. Procedure

2-(2-Methoxyphenyl)-2-methylpropionitrile.   A 300-mL, three-necked, round-bottomed flask equipped with a reflux condenser and a Teflon-coated magnetic stir bar is placed under a nitrogen atmosphere and charged with 2-fluoroanisole (10.00 g, 79.28 mmol) (Note 1), 100 mL of tetrahydrofuran   (THF, Note 2) and potassium hexamethyldisilylamide (KHMDS, 23.72 g, 118.92 mmol) (Note 3). Isobutyronitrile (28.8 mL, 316.71 mmol) (Note 4) is added via syringe. The reaction mixture is heated to 60°C for 23 hr (Note 5). After the solution is cooled to room temperature, it is transferred to a 1000-mL separatory funnel containing 300 mL of methyl tert-butyl ether (Note 6) and 300 mL of 1 N hydrochloric acid (HCl). The organic layer is separated and washed successively with 300 mL of water and 200 mL of brine. The organic phase is dried over anhydrous magnesium sulfate (MgSO$_4$), filtered, and concentrated to provide the desired product as a tan oil (13.75 g, 99%) (Notes 7, 8 and 9).

2. Notes

1. 2-Fluoroanisole was purchased from Aldrich Chemical Co., Inc., and used without further purification.

2. Anhydrous tetrahydrofuran was purchased from Aldrich Chemical Co., Inc., in a Sure/Seal™ bottle.

3. Potassium hexamethyldisilylamide was purchased from Aldrich Chemical Co., Inc., and used without further purification. The solid was weighed on a balance without special protection from air.

4 Isobutyronitrile was purchased from Aldrich Chemical Co., Inc., and used without further purification.

5. The reaction can be monitored by HPLC (Hewlett-Packard 1100 HPLC, Kromasil C18 column (4.6 x 150 mm), 50/50 MeCN/0.2% H$_3$PO$_4$, 1.0 mL/min, product = 7.8 min) or by TLC analysis (2-fluoroanisole R$_f$ = 0.67, product R$_f$ = 0.51, ethyl acetate/hexanes 20/80).

6. A.C.S. Reagent grade methyl tert-butyl ether was purchased from J. T. Baker and used as received.

7. ^{1}H NMR and ^{13}C NMR indicate reasonably pure product. The HPLC analysis of the crude product showed a purity >97% (same conditions as Note 5) and satisfactory elementary analysis (calculated for C$_{11}$H$_{13}$NO: C, 75.40; H, 7.48; N, 7.99. Found: C, 75.01; H, 7.35; N, 8.07).

8. The product shows the following physical properties: ^{1}H NMR (CDCl$_3$, 300 MHz) δ: 1.80 (s, 6), 3.96 (s, 3), 6.97-7.02 (m, 2), 7.29-7.39 (m, 2); ^{13}C NMR δ: 27.00,

34.43, 55.51, 112.02, 120.76, 124.80, 125.92, 128.62, 129.39, 157.30. IR cm^{-1}: 2980, 2235, 1493, 1462, 1437, 1253, 1027, 756.

9. If material of greater purity is necessary, the product can be purified by chromatography on silica gel (100 g) using ethyl acetate/hexanes 20/80 (900 mL) as the eluant. When 100-mL fractions were collected, fractions #2 to #5 contained the desired product in >99% purity (12.31g, 89% yield)

Waste Disposal Information

All toxic materials were disposed of in accordance with "Prudent Practices in the Laboratory;" National Academy Press; Washington, DC, 1995.

3. Discussion

Tertiary benzylic nitriles are useful synthetic intermediates, and have been used for the preparation of amidines,[2] lactones,[3] primary amines,[4,5] pyridines,[6] aldehydes,[7,8] carboxylic acids,[9] and esters.[10,11] The general synthetic pathway to this class of compounds relies on the displacement of an activated benzylic alcohol or benzylic halide with a cyanide source followed by double alkylation under basic conditions. For instance, 2-(2-methoxyphenyl)-2-methylpropionitrile has been prepared by methylation of (2-methoxyphenyl)acetonitrile using sodium amide and iodomethane.[12] In the course of the preparation of a drug candidate,[13] the submitters discovered that the nucleophilic aromatic substitution of aryl fluorides with the anion of a secondary nitrile is an effective method for the preparation of these compounds.[14] The reaction was studied using isobutyronitrile and 2-fluoroanisole. The submitters first showed that KHMDS was the superior base for the process when carried out in either THF or toluene (Table I). For example, they found that the preparation of 2-(2-methoxyphenyl)-2-methylpropionitrile could be accomplished in

211

near quantitative yield in a single operation. Several other substrates were studied as summarized in Table II. The reaction proceeds using either electron-rich or electron-poor arenes (entries 1-11) as well as cyclic nitriles (entries 12-14). The reaction requires an excess of nitrile, which can self-condense under the reaction conditions, as well as a slight excess of base. The submitters obtained the best results when using about 1.5 equiv of base in conjunction with 4 equiv of the nitrile. The acidic work-up removes most of the impurities generated in the reaction and, in the case of a low boiling starting nitrile, the excess reagent is evaporated with the solvent to provide a crude material of high purity.

1. Process Research, Chemical Research and Development, Pfizer Global R&D, Groton, CT 06340-8156. E. Vazquez' current address is Merck & Co. Inc., RY-800-C367, P.O. Box 2000, Rahway NJ 07065-0900.

2. Convery, M. A.; Davis, A. P.; Dunne, C. J.; MacKinnon, J. W. *Tetrahedron Lett.* **1995**, *36*, 4279-4282.

3. Tiecco, M.; Testaferri, L.; Tingoli, M.; Bartoli, D. *Tetrahedron* **1990**, *46*, 7139-7150.

4. O'Donnell, M. J.; Wu, S.; Huffman, J. C. *Tetrahedron* **1994**, *50*, 4507-4518.

5. Okatani, T.; Koyama, J.; Tagahara, K. *Heterocycles* **1989**, *29*, 1809-1814.

6. Chelucci, G.; Conti, S.; Falorni, M.; Giacomelli, G. *Tetrahedron* **1991**, *47*, 8251-8258.

7. Cha, J. K.; Christ, W. J.; Kishi, Y. *Tetrahedron Lett.* **1983**, *24*, 3943-3946.

8. Chavan, S. P.; Ravindranathan, T.; Patil, S. S.; Dhondge, V. D.; Dantale, S. W. *Tetrahedron Lett.* **1996**, *37*, 2629-2630.

9. Leader, H.; Smejkal, R. M.; Payne, C. S.; Padilla, F. N.; Doctor, B. P.; Gordon, R. K.; Chiang, P. K. *J. Med. Chem.* **1989**, *32*, 1522-1528.

10. Bush, E. J.; Jones, D. W. *J. Chem. Soc., Perkin Trans. 1* **1997**, 3531-3536.

11. Breukelman, S. P.; Meakins, G. D.; Roe, A. M. *J. Chem. Soc., Perkin Trans. 1* **1985**, 1627-1635.

12. Gripenberg, J.; Hase, T. *Acta Chem. Scand.* **1966**, *20,* 1561-1570.

13. Caron, S. In *Organic Reactions and Processes Gordon Research Conference 2000*; Roger Williams University, 2000.

14. Caron, S.; Vazquez, E.; Wojcik, J. M. *J. Am. Chem. Soc.* **2000**, *122*, 712-713.

TABLE I

NUCLEOPHILIC AROMATIC SUBSTITUTION OF (2-METHOXYPHENYL)-
ACETONITRILE WITH ISOBUTYRONITRILE

Entry	Base (1.5 equiv)	Solvent	T (°C)	Time (h)	Yield[a] (%)
1	LiHMDS	THF	60	23	3
2	NaHMDS	THF	60	23	49
3	KHMDS	THF	60	23	95
4	KHMDS	Toluene	60	18	95
5	KHMDS	DMSO	75	24	No rxn
6	KHMDS	i-Pr$_2$O	75	24	3
7	KHMDS	NMP	75	24	1

a) Yields <5% indicates the conversion observed by HPLC analysis after the time shown. The yields of entry 2,3 and 4 are isolated yields.

TABLE II

S_NAr OF ARYL FLUORIDES WITH SECONDARY NITRILES AND KHMDS

Entry	Fluoride	Nitrile (equiv)	Solvent	T (°C)	Time	Yield (%)
1	2-OMe	**1** (4.0)	Toluene	60	18 h	99
2	3-OMe	**1** (4.1)	Toluene	100	3 h	69
3	4-OMe	**1** (4.0)	THF	60	50 h	66
4	3,5-OMe	**1** (4.0)	Toluene	70	48 h	85
5	3,4-OMe	**1** (4.0)	THF	80	4 h	28
6	2-Cl	**1** (4.0)	Toluene	60	45 min	77
7	4-Cl	**1** (4.0)	THF	75	2 h	72
8	H	**1** (4.0)	THF	75	14 h	69
9	3-Me	**1** (4.0)	THF	75	14 h	71
10	4-CN	**1** (4.0)	Toluene	60	40 min	83
11	4-CF$_3$	**1** (3.3)	Toluene	60	40 min	94
12	2-OMe	**2** (4.0)	Toluene	75	48 h	47
13	2-OMe	**3** (4.0)	THF	75	24 h	67[a]
14	2-OMe	**4** (4.0)	THF	75	15 h	70[a]

a) Provided exclusively the 2(*S*) isomer (*exo*-aryl).

Appendix

Chemical Abstracts Nomenclature (Collective Index Number); (Registry Number)

2-(2-Methoxyphenyl)-2-methylpropionitrile: Hydratroponitrile, o-methoxy-α-methyl- (8); Benzeneacetonitrile, 2-methoxy-α,α-dimethyl- (9); (13524-75-9)

2-Fluoroanisole: Anisole, o-fluoro- (8); Benzene, 1-fluoro-2-methoxy- (9); (321-28-8)

Potassium hexamethyldisilylamide: Aldrich: Potassium-(bis(trimethylsilyl)amide: Silanamine, 1,1,1-trimethyl-N-(trimethylsilyl)-, potassium salt (9); (40949-94-8)

Isobutyronitrile (8): Propanenitrile, 2-methyl- (9); (78-82-0)

PREPARATION OF 1-[N-BENZYLOXYCARBONYL-(1S)-1-AMINO-2-OXOETHYL]-4-METHYL-2,6,7-TRIOXABICYCLO[2.2.2]OCTANE

[Carbamic acid, [1-(4-methyl-2,6,7-trioxabicyclo[2.2.2]oct-1-yl)-2-oxoethyl]-, phenylmethyl ester, (S)-]

Submitted by Nicholas G. W. Rose, Mark A. Blaskovich, Ghotas Evindar,
Scott Wilkinson, Yue Luo, Dan Fishlock, Chris Reid, and Gilles A. Lajoie.[1]
Checked by Richard S. Gordon and Andrew B. Holmes

1. Procedure

A. *3-Methyl-3-(toluenesulfonyloxymethyl)oxetane [Oxetane tosylate, (1)]*. A dry, 1-L, round-bottomed flask is charged with p-toluenesulfonyl chloride (57.20 g, 0.30 mol) (Note 1) to which pyridine (250 mL) (Note 2) is added while stirring is carried out under nitrogen with a magnetic stir bar (Note 3). The reaction flask is placed inside a container to which an ice/water mixture may be added in the event that the reaction becomes too exothermic. 3-Methyl-3-oxetanemethanol (20.4 g, 0.2 mol) (Note 4) is added slowly and stirred for 1.5 hr. The mixture is slowly added to a vigorously (Note 3) stirred mixture of de-ionized water (700 mL) and crushed ice (700 g) in a 2-L Erlenmeyer flask and allowed to stir for an additional 0.5 hr. The white precipitate is collected on Whatman filter paper # 1 and washed with cold water (H_2O, Note 5). The product is dried under high vacuum and/or phosphorus pentoxide (P_2O_5) to obtain the white powder of oxetane tosylate 1 (39.8-44.60 g, 78-87%) (Note 6).

B. *N-Benzyloxycarbonyl-L-serine 3-methyl-3-(hydroxymethyl)oxetane ester [Cbz-Ser oxetane ester, (2)]*. Cbz-L-Ser (11.36 g, 0.047 mol) (Note 7) and cesium carbonate (Cs_2CO_3, 9.19 g, 0.028 mol, 0.6 equiv) (Note 8) are combined in a 500-mL, round-bottomed flask and dissolved in H_2O (100 mL). The water is removed under reduced pressure and the resulting oil is lyophilized for 12 hr to give a white foam. To this foam are added oxetane tosylate 1 (12.65 g, 0.049 mol) and sodium iodide (NaI, 1.41 g, 9.8 mmol, 0.2 equiv) (Note 9), which is taken up in dimethylformanide (DMF, 400 mL) (Note 10) and the solution is stirred with a magnetic stir bar at 25°C under argon (Ar) for 48 hr. The DMF is removed under reduced pressure (0.5 mm, bath temperature 50°C) and the resulting solid is dissolved in a two-phase mixture by sequential treatment with alternating aliquots of ethyl acetate (EtOAc, 600 mL total) and H_2O (200 mL total). The organic phase is separated and washed with aqueous 10% sodium bicarbonate (NaHCO$_3$, 2 x 100 mL) (Note 11) and saturated aqueous sodium chloride (NaCl, 1 x 100 mL), dried over magnesium sulfate (MgSO₄), and evaporated on a rotary evaporator. The solvent is removed under reduced pressure to yield a yellow oil that is recrystallized from diethyl ether to yield colorless, rod-like crystals in 70-72% yield (10.6-10.9 g) (Note 12).

C. *1-[N-Benzyloxycarbonyl-(1S)-1-amino-2-hydroxyethyl]-4-methyl-2,6,7-trioxa-bicyclo[2.2.2]octane [Cbz-L-Ser OBO ester, (3)].* In a 500-mL, round-bottomed flask, Cbz-Ser oxetane ester **2** (11.85 g, 36.6 mmol) (Note 13) is dissolved in dry dichloromethane (CH₂Cl₂, 450 mL) (Note 14) and cooled to 0°C under Ar. Boron trifluoride etherate (BF₃·Et₂O, 0.23 mL, 1.83 mmol) (Note 15) is diluted in dry CH₂Cl₂ (5.0 mL) and added by syringe to the reaction flask. The solution is stirred with a magnetic stir bar at 25°C under Ar for 6 hr, at which point TLC indicates that the reaction is over. Triethylamine (Et₃N, 1.28 mL, 9.15 mmol) (Note 16) is added and the reaction is stirred for an additional 30 min before being concentrated to a thick oil. The crude product is redissolved in EtOAc (400 mL) and washed with aqueous 3% ammonium chloride (NH₄Cl, 2 x 250 mL) (Note 17), aqueous 10% NaHCO₃ (1 x 250 mL), saturated aqueous NaCl (1 x 250 mL), dried (MgSO₄), and evaporated to dryness. The reaction yields a colorless thick oil, (14.2 g, 95% yield). The clear colorless oil is crystallized from EtOAc/hexane to give (9.7-11 g, 82-93% yield) of shiny crystals (Notes 18 - 20).

D. *1-[N-Benzyloxycarbonyl-(1S)-1-amino-2-oxoethyl]-4-methyl-2,6,7-trioxabi-cyclo[2.2.2]octane, [Cbz-L-Ser(ald) OBO ester, (4)].* Cbz-Ser OBO ester **3** (9.10 g, 28.0 mmol) (Note 21) is dissolved in dry CH₂Cl₂ (80 mL) (Note 14) under Ar and cooled to -78°C in a 100-mL, round-bottomed flask labeled flask 1. Oxalyl chloride (3.9 mL, 45 mmol, 1.61 equiv) (Note 22) is added to dry CH₂Cl₂ (120 mL) (Note 14) in a separate 250-mL, round-bottomed flask (flask 2) under Ar, and cooled to -78°C. Dry dimethyl sulfoxide (DMSO, 7.0 mL, 90 mmol, 3.21 equiv) (Note 23) is added to the oxalyl chloride solution (flask 2) and the mixture is stirred under Ar (magnetic stir bar) at -78°C for 15 min. The alcohol solution **3** (in flask 1) is transferred slowly by cannula to flask 2 over a period of 45 min and then rinsed with dry CH₂Cl₂ (50 mL) (Note 14). The resulting cloudy white mixture is stirred for 1.5 hr at -78°C. Diisopropylethylamine (DIPEA, 24.27 mL, 0.14 mol, 5.0 equiv) (Note 24) is added and the solution is stirred for 30 min at -78°C and 10 min at 0°C. Ice-cold CH_2Cl_2 (250 mL) is added and the solution is washed with ice-cold aqueous 3% NH_4Cl (3 x 250 mL), aqueous 10% $NaHCO_3$ (1 x 250 mL), saturated aqueous NaCl (1 x 250 mL), dried ($MgSO_4$), and evaporated to dryness on a rotary evaporator. The reaction yields 8.4-8.77 g (92-96%) of a slightly yellowish solid that may be used without further purification (Notes 25, 26, 27).

2. Notes

1. p-Toluenesulfonyl chloride (>97%) was purchased from Aldrich Chemical Company, Inc., (checkers) or Fluka Chemicals (submitters) and used without further purification.

2. Pyridine (>99%) was purchased from Fisher Scientific, UK (checkers) or Fisher Scientific Company (submitters) and used without further purification.

3. The checkers found that vigorous stirring was essential.

4. 3-Methyl-3-oxetanemethanol (98%) was purchased from Aldrich Chemical Company, Inc., and used without further purification.

5. The checkers cooled 250 mL of distilled water in a refrigerator for this purpose.

6. The submitters obtained 49.1 g (96%). The product has the following physical and spectral characteristics: mp 59-60°C or 61-62°C; the submitters observed mp 49.5-51°C; (lit.[2] mp 49.5-51°C); TLC (3:2 v/v, hexane:ethyl acetate) R_f (Merck kieselgel) = 0.42; ^{1}H NMR ($CDCl_3$, 250 MHz) δ: 1.31 (s, 3 H), 2.46 (s, 3 H), 4.11 (s, 2 H), 4.35 (d, 2 H, J = 6.3), 4.35 (d, 2 H, J = 6.3), 7.37 (d, 2 H, J = 8.2), 7.81 (d, 2 H, J = 8.2); ^{13}C NMR ($CDCl_3$, 101 MHz) δ: 20.6, 21.6, 39.3, 74.3, 78.9, 128.0, 130.0, 132.7, 145.1; IR (cast from $CHCl_3$) cm[-1]: 2958, 2877, 1531, 1364, 1226, 1223, 1189, 1177; HRMS (ES, M +

Na$^+$) m/z Calcd for $C_{12}H_{16}O_4Na$: 279.0667. Found: 279.0656; Anal. Calcd for $C_{12}H_{16}$ O_4S: C, 56.2; H, 6.3; N. Found: C, 56.3; H, 6.4.

7. The checkers purchased Cbz-L-Serine from Nova Biochem while the submitters purchased this from Advanced Chemtech; in both cases the material was used without further purification.

8. Cesium carbonate (99%) was purchased from Aldrich Chemical Company, Inc., and used without further purification.

9. Sodium iodide (99%) was purchased from Aldrich Chemical Company, Inc., and used without further purification.

10. DMF (99.8%) was purchased from BDH and stored over activated 4Å molecular sieves (8-12 Mesh, purchased from Acros Organics) before use.

11. The NaHCO$_3$ solution was prepared and used immediately.

12. The submitters recrystallized the product from ethyl acetate and hexanes to obtain a 78% yield. The checkers obtained crude product **2** in 85-94% yield. They found the recrystallization of **2** difficult and preferred diethyl ether as solvent. The crude product was initially dissolved in diethyl ether (ca. 50 mL) and left open to the atmosphere to reduce the volume to about 15 mL. The yield is based on the recovery of solid from two crops. The checkers found that the use of diethyl ether, although time consuming, was a more reliable procedure for recrystallization. The chemical properties of **2** are as follows: mp 58-60°C (Et$_2$O) (69-71°C from EtOAc-hexane) (submitters and lit.[2] 70-70.5°C from EtOAc-hexanes); sample recrystallized from Et$_2$O $[\alpha]_D^{20}$-8.6° (EtOAc, c 1.0) sample recrystallized from EtOAc-hexanes $[\alpha]_D^{20}$-8.3° (EtOAc, c 1.0); the submitters reported $[\alpha]_D^{20}$-8.5° (EtOAc, c 1.04); TLC (2:1, EtOAc:hexane), R$_f$ = 0.34; ^{1}H NMR (CDCl$_3$, 250 MHz) δ: 1.28 (br s, 3 H), 3.01 (t, 1 H, J = 6.0), 3.80-3.93 (br m, 1 H), 4.04-4.13 (br m, 2 H), 4.38-4.56 (m, 6 H), 5.12 (s, 2 H), 5.89 (d, 1 H, J = 7.9), 7.30-7.40 (br m, 5 H); ^{13}C NMR (CDCl$_3$, 125 MHz) δ: 20.7, 39.6, 56.4, 63.3, 67.1, 68.9, 79.4, 128.1, 128.2, 128.5, 136.1, 156.2, 170.7; IR (cast from CHCl$_3$) cm^{-1}: 3329, 2958, 2877, 1714, 1527, 1214,

1062, 976, 752; HRMS (ES, M + Na$^+$) m/z Calcd for C$_{16}$H$_{21}$NO$_6$Na: 346.1267. Found: 346.1257. Anal. Calcd for C$_{16}$H$_{21}$NO$_6$: C, 59.4; H, 6.6; N, 4.3. Found: C, 59.5; H, 6.6; N, 4.4.

13. The checkers obtained the required quantity of ester **2** by combining the product from two runs in Step B. Alternatively, in several trial experiments, the checkers used the crude ester **2** as a starting material with no appreciable decrease in yield of compound **3**.

14. Dichloromethane is freshly distilled from CaH$_2$.

15. Boron trifluoride diethyl etherate (redistilled) was purchased from Aldrich Chemical Company, Inc., and used without further purification.

16. Triethylamine (99%) was purchased from Aldrich Chemical Company, Inc., and used without further purification.

17. The checkers found the work-up lengthy as the phases took extended periods to separate (15 min). Significant emulsion formation (with product loss) on work-up was observed if aqueous solutions were marginally more concentrated (see Note 11).

18. Care must be taken not to expose product **3** to aqueous acid conditions for prolonged periods of time since ring opening of the OBO will occur. Thus care must be taken upon both addition of BF$_3$·Et$_2$O to **2** and upon work-up, hence the sodium bicarbonate wash after the 3% NH$_4$Cl extraction. The diol that is formed upon acid-catalyzed hydrolysis of the OBO also crystallizes out making purification of the desired product **3** impossible.

19. The checkers also prepared racemic 1-[N-benzyloxycarbonyl-(1±)-1-amino-2-hydroxyethyl]-4-methyl-2,6,7-trioxabicyclo[2.2.2]octane using the identical procedure (50% over 2 steps from racemic Cbz-serine purchased from Aldrich Chemical Company, Inc.). The enantiomeric ratio of the crystalline (S)-**3** enantiomer (Note 20) was > 99.5:0.5 as determined by comparison with racemic **3** by courtesy of Mr. Eric Hortense (GlaxoSmithKline, Stevenage). Chiral HPLC (25 cm Chiracel OD-H, Column No ODHOCE-IF029, mobile phase ethanol/heptane 1:4 v/v, UV detector at 215 nm, flow rate

1.0 mL/min at room temperature) afforded the (S)-**3** enantiomer with a retention time of 8.1 min while the (R)-**3** enantiomer had a retention time of 11.0 min.

20. The submitters recrystallized the sample from EtOAc-hexanes and obtained a yield of 93%. The checkers' yield is based on combined product from two successive crops, although additional product was observed in the remaining mother-liquor (2 g of crude material) which could be used, as obtained, for subsequent reactions. The physical and spectroscopic properties of **3** are as follows: mp 104-106°C (material isolated b y column chromatography alone); mp 110-112°C $[\alpha]_D^{20}$-24.6° (EtOAc, c 1.0) (after a single crystallization from EtOAc/hexane); mp 119-121°C $[\alpha]_D^{20}$-24.1° (EtOAc, c 0.8) (after two recrystallizations) with all samples exhibiting analytical purity by TLC analysis; the submitters reported mp 103.5-105°C; $[\alpha]_D^{20}$-24.8° (EtOAc, c 1.00); TLC (3:1 EtOAc:hexane), R_f = 0.37; ^{1}H NMR (CDCl$_3$, 250 MHz) δ: 0.81 (s, 3 H), 2.57 (m, 1 H), 3.61-3.95 (m, 9 H), 5.10-5.18 (m, 2 H), 5.33 (d, 1 H, J = 8.8), 7.29-7.38 (m, 5 H); ^{13}C NMR (CDCl$_3$, 101 MHz) δ: 14.2, 30.5, 55.2, 61.9, 66.9, 72.7, 108.4, 128.1, 128.2, 128.5, 136.4, 156.3; IR (cast from CHCl$_3$) cm^{-1}: 3019, 2966, 2881, 1717, 1519, 1216; HRMS (ES, M + Na$^+$) m/z Calcd for C$_{16}$H$_{21}$NO$_6$Na: 346.1267. Found: 346.1260. Anal. Calcd for C$_{16}$H$_{21}$NO$_6$: C, 59.4; H, 6.55; N, 4.3. Found: C, 59.2; H, 6.6; N, 4.2. Chiral HPLC of **3** displayed a single enantiomer (R$_f$=8.1 min) (Note 19). It is essential that CDCl$_3$ used for NMR samples containing the ortho ester be prefiltered through basic alumina to remove traces of acid.

21. The checkers obtained the required quantity of alcohol **3** by combining the product from two runs in step C.

22. Oxalyl chloride (98%) was purchased from Aldrich Chemical Company, Inc., and used without further purification. Amounts of reagents for the Swern oxidation have been optimized as reported. Alternative amounts reduce both yield and % ee.

23. DMSO (dimethyl sulfoxide) (99.8%) stored in an Aldrich Sure/Seal bottle was purchased from Aldrich Chemical Company, Inc., and used without further purification.

24. DIPEA (diisopropylethylamine, redistilled 99.5%) was purchased from Aldrich Chemical Company, Inc., and used without further purification.

25. The submitters obtained 8.68 g (96%). The aldehyde **4** should be used immediately after preparation. It cannot be purified by chromatography. The checkers observed the following physical and spectroscopic properties of **4**: $[\alpha]_D^{20}$ (EtOAc, c 1.0) fell in the range -36° to -62.0° and was considered an unreliable estimate of enantiomeric purity (see Note 26); the submitters obtained $[\alpha]_D^{20}$-99.3° (EtOAc, c 1.03) (lit. 8); TLC (3:1 EtOAc:hexane), R_f = 0.60; ^{1}H NMR (CDCl$_3$, 250 MHz) δ: 0.83 (s, 3 H), 3.94 (s, 6 H), 4.60 (d, 1 H, J = 8.9), 5.08-5.14 (m, 2 H), 5.38 (d, 1 H, J = 9.2), 7.30-7.38 (m, 5 H), 9.69 (s, 1 H); ^{13}C NMR (CDCl$_3$, 101 MHz) δ: 14.2, 30.9, 63.3, 67.2, 72.9, 107.2, 128.1, 128.5, 136.2, 156.2, 195.6; IR (cast from CHCl$_3$) cm^{-1}: 2947, 2883, 1723(br), 1517, 1218. HRMS (ES, M + Na$^+$) m/z Calcd for C$_{16}$H$_{19}$NO$_6$Na: 344.1440. Found: 346.1106. Anal. Calcd for C$_{16}$H$_{19}$NO$_6$: C, 59.75; H, 6.0; N, 4.4. Found: C, 59.4; H, 6.1; N, 4.3.

26. To confirm the enantiomeric integrity of aldehyde **4**, and in view of the variability of the specific rotation, the checkers reduced aldehyde **4** to the alcohol **3**. Aldehyde **4** (6.9 g, 21.3 mmol) was dissolved in THF/EtOH (60 mL, 1:1), and the solution was stirred with a magnetic stir bar in a 100-mL round-bottomed flask at -20°C. Sodium borohydride (0.8 g, 22.2 mmol) (Note 26) was added as a suspension in water (2 mL), and the mixture was stirred for 1 hr. The organic solvents were removed and the residue was subsequently redissolved in EtOAc (100 mL). The organic phase was washed with aqueous 3% NH$_4$Cl (2 x 100 mL), aqueous 10% NaHCO$_3$ (1 x 100 mL) and saturated aqueous NaCl solution. This was dried (MgSO$_4$) and the solvent was removed (rotary evaporator) to afford alcohol **3** (6.5 g, 20.2 mmol, 95%) as a white solid. This was recrystallized as described in Step C (6.0 g). Chiral HPLC analysis of the recrystallized product, under the previously described conditions (Note 19), showed **3** having an enantiomeric ratio 99.5:0.5. The mother-liquors (0.5 g) from the crystallization of **3** contained an enantiomeric ratio > 85:15 of the (S)-and (R)-enantiomers respectively. These figures would correspond to an enantiomeric ratio of

98.4:1.6 for the as prepared Cbz-L-Ser(ald) OBO ester **4**, assuming no loss of material. The submitters determined the enantiomeric purity of Cbz-Ser(ald) OBO ester by chiral shift ^{1}H NMR studies. Cbz-Ser(ald) OBO ester **4** (10 mg) was dissolved in benzene-d_6. Eu(hfc)$_3$ (100 μL, 50 mg/mL in benzene-d_6) was added to obtain the ^{1}H NMR spectrum at 250 MHz. The purity was observed to be 97-99% ee.

27. Sodium borohydride was purchased from Avocado, and used without purification.

Waste Disposal Information

All toxic materials were disposed of in accordance with "Prudent Practices in the Laboratory"; National Academy Press; Washington, DC, 1995.

3. Discussion

The nonproteinogenic α-amino acids represent an important group of natural products because of their varied biological properties.[2] One approach to the synthesis of these compounds is through elaboration of the chiral pool of 19 natural α-amino acids. Serine aldehyde derivatives are perhaps the most popular chiral synthons for the synthesis of nonproteinogenic α-amino acids; the Garner aldehyde has been especially popular in that regard.[3] The submitters have developed an alternative chiral serine aldehyde synthon **4** that is both more facile in preparation and prepared in higher yields (70% over 3 steps).[4] However, the greatest benefit exists in the fact that conversion to the free α-amino acid proceeds through deprotection of the intermediate, unlike Garner's aldehyde that requires further oxidation to achieve the α-amino acid, and may suffer from racemization.[5] The conversion of the α-carboxylic acid to the OBO protecting group reduces the acidity of the α-proton, preventing racemization in both olefination[6] and nucleophilic[4,7-9] addition

224

homologation reactions to the aldehyde **4**. The procedure described here provides aldehyde **4** determined to be 97-99% enantiomerically pure by chiral shift ^{1}H NMR studies and HPLC analysis.[7] Conversion to the free amino acid can be achieved by acid hydrolysis to yield products typically >98% enantiomerically pure.[4] Examples of the use of 1-[N-benzyloxycarbonyl-(1S)-1-amino-2-oxoethyl]-4-methyl-2,6,7-trioxabicyclo[2.2.2] oxetane **4** as a chiral synthon for natural product synthesis are shown in Scheme 1.

Scheme 1

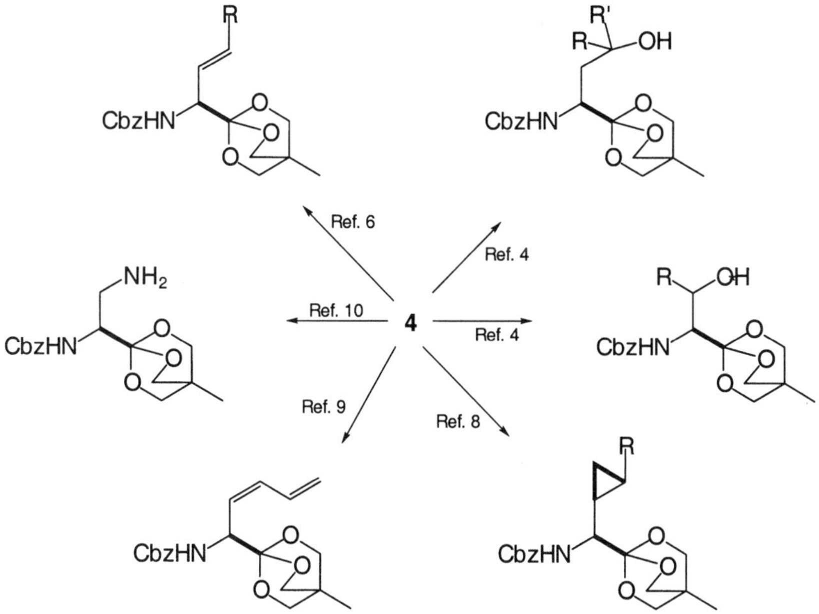

1. Department of Chemistry, University of Waterloo, Waterloo, Ontario, Canada, N2L 3G1.
2. Williams, R. M. "Synthesis of Optically Active α-Amino Acids"; Pergamon Press: New York, 1989.

3. (a) Garner, P. *Tetrahedron Lett.* **1984**, *25*, 5855; (b) Garner, P.; Park, J. M. *J. Org. Chem.* **1990**, *55*, 3772.

4. Blaskovich, M. A.; Evindar, G.; Rose, N. G. W.; Wilkinson, S.; Luo, Y.; Lajoie, G. A. *J. Org. Chem.* **1998**, *63*, 3631 and 4560.

5. Beaulieu, P. L.; Duceppe, J.-S.; Johnson, C. *J. Org.Chem.* **1991**, *56,* 4196.

6. Rose, N. G. W.; Blaskovich, M. A.; Wong, A.; Lajoie, G. A. *Tetrahedron*, **2001**, *57*, 1497-1507.

7. Blaskovich, M. A.; Lajoie, G. A. *J. Am. Chem. Soc.* **1993**, *115*, 5021.

8. Rifé, J.; Ortuño, R. M.; Lajoie, G. A. *J. Org. Chem.* **1999**, *64*, 8958.

9. Cameron, S.; Khambay, B. P. S. *Tetrahedron Lett.* **1998**, *39*, 1987.

10. Luo, Y.; Blaskovich, M. A.; Lajoie, G. A. *J. Org. Chem.* **1999**, *64*, 6106.

Appendix
Chemical Abstracts Nomenclature (Collective Index Number);
(Registry Number)

1-[N-Benzyloxycarbonyl-(1S)-1-amino-2-oxoethyl]-4-methyl-2,6,7-trioxabicyclo[2.2.2]octane: Carbamic acid, [1-(4-methyl-2,6,7-trioxabicyclo[2.2.2]oct-1-yl)-2-oxoethyl]-, phenylmethyl ester, (S)- (14); (183671-34-3)

3-Methyl-3-(toluenesulfonyloxymethyl)oxetane: 3-Oxetanemethanol, 3-methyl-, 4-methylbenzenesulfonate (11); (99314-44-0)

p-Toluenesulfonyl chloride (8); Benzenesulfonyl chloride, 4-methyl- (9); (98-59-9)

Pyridine (8, 9); (110-86-1)

3-Methyl-3-oxetanemethanol: 3-Oxetanemethanol, 3-methyl- (9); (3143-02-0)

N-Benzyloxycarbonyl-L-serine 3-methyl-3-(hydroxymethyl)oxetane ester: L-Serine, N-[(phenylmethoxy)carbonyl]-, (3-methyl-3-oxetanyl)methyl ester (14); (206191-42-6)

Cesium carbonate: Carbonic acid, dicesium salt (8, 9); (534-17-8)

N-(Benzyloxycarbonyl)-L-serine: L-Serine, N-[(phenylmethoxy)carbonyl]- (9); (1145-80-8)

Sodium iodide (8, 9); (7681-82-5)

N,N-Dimethylformanide: CANCER SUSPECT AGENT: Formamide, N,N-dimethyl- (8, 9); (68-12-2)

1-[N-Benzyloxycarbonyl-(1S)-1-amino-2-hydroxyethyl]-4-methyl-2,6,7,trioxabicyclo[2.2.2]octane: Carbonic acid, [(1S)-2-hydroxy-1-(4-methyl-2,6,7-trioxabicyclo[2.2.2]oct-1-yl)ethyl]-, phenylmethyl ester (14); (206191-44-8)

Boron trifluoride etherate: Ethyl ether, compd. with boron fluoride (BF_3) (1:1) (8); Ethane, 1,1'-oxybis-, compd. with trifluoroborane (1:1) (9); (109-63-7)

Triethylamine (8); Ethanamine, N,N-diethyl- (9); (121-44-8)

Oxalyl chloride: HIGHLY TOXIC: (8); Ethanedioyl dichloride (9); (79-37-8)

Dimethyl sulfoxide: Methyl sulfoxide (8); Methane, sulfinylbis- (9); (67-68-5)

N,N-Diisopropylethylamine: Triethylamine, 1,1'-dimethyl- (8); 2-Propanamine, N-ethyl-N-(1-methylethyl)- (9); (7087-68-5)

N-HYDROXY-4-(p-CHLOROPHENYL)THIAZOLE-2(3H)-THIONE

[2(3H)-Thiazolethione, 4-(4-chlorophenyl)-3-hydroxy-]

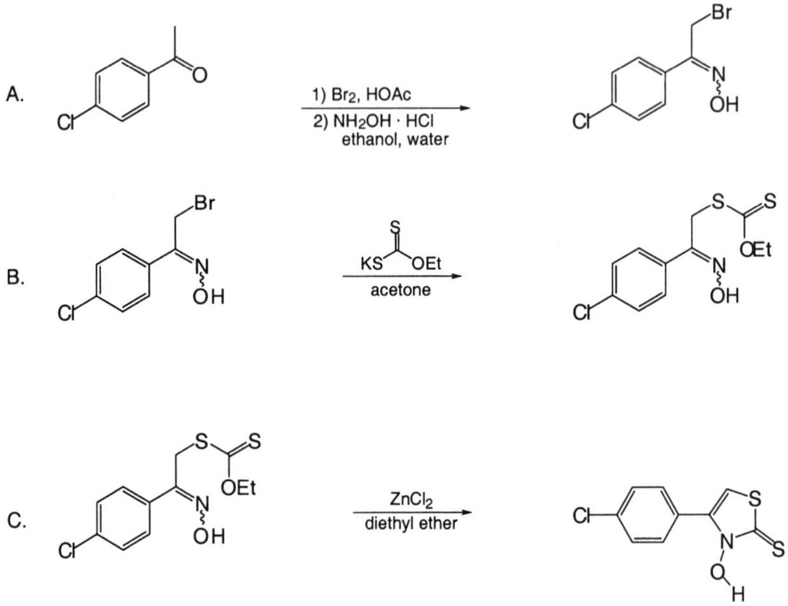

A.

1) Br₂, HOAc
2) NH₂OH · HCl
ethanol, water

B.

KS OEt

acetone

C.

ZnCl₂
diethyl ether

Submitted by Jens Hartung and Michaela Schwarz.[1]

Checked by Raghuram S. Tangirala and Dennis P. Curran.

1. Procedure

A. ω-Bromo-p-chloroacetophenone oxime. A 500-mL, four-necked, round-bottomed flask equipped with a dropping funnel (closed with a glass stopper), mechanical stirrer, drying tube (calcium chloride, $CaCl_2$), and a thermometer is charged with p-chloroacetophenone (38.6 g, 0.25 mol), glacial acetic acid (220 mL) and aqueous hydrobromic acid (HBr) [1 mL, 48 % (w/w), Note 1]. The flask is immersed in a water bath (15°C, Note 2) and bromine (40.0 g, 0.25 mol) is added from the dropping funnel at such a rate that the temperature of the reaction mixture does not exceed 25°C (Note 3). Stirring is continued for 1 hr at 15°C. The suspension is poured onto 1 kg of crushed ice. The colorless precipitate is collected by suction (Buchner funnel) and repeatedly washed with small portions of water (total of 200 mL). Approximately 84 g of air-dried ω-bromo-p-chloroacetophenone is obtained from this step (Note 4). The crude product is transferred into a 1-L beaker that is equipped with a magnetic stirring bar (Note 5). Aqueous ethanol (96%, v/v, 400 mL) is added and the slurry is treated with a solution of hydroxylamine hydrochloride [21.7 g, 0.31 mol, (Note 6)] in water (65 mL) in one portion at 20°C. Stirring is continued for 24 hr at 20°C to afford a clear, pale yellow solution that is poured onto a mixture of ice (900 g) and water (400 mL). ω-Bromo-p-chloroacetophenone oxime separates as a colorless solid that is collected by filtration on a Buchner funnel. The precipitate is washed with small portions of water (total of 200 mL) and dried for 48 hr at 20°C and 10^{-3} mbar (7.5 x 10^{-4} mm) to furnish 53 g (85.3%) of ω-bromo-p-chloroacetophenone oxime as a colorless powder (Note 7).

B.   *O-Ethyl S-[oximino-2-(p-chlorophenyl)ethyl]dithiocarbonate.*   A 1-L, round-bottomed flask equipped with a dropping funnel is charged with a magnetic stirring bar, potassium O-ethyl xanthate (35.3 g, 0.22 mol), and acetone (140 mL). A solution of ω-bromo-p-chloroacetophenone oxime (49.5 g, 0.20 mol) in acetone (120 mL) is added dropwise at room temperature over a period of 30 min. Stirring is continued for 3 hr at 20°C whereupon a cloudy orange solution forms from the initial slurry. Solids are removed by filtration using a thin pad of diatomaceous earth (1-cm height in a Buchner funnel) to afford a clear orange solution. The solids on the funnel are washed with acetone (total of 50 mL). The combined washings and filtrate are concentrated under an aspirator vacuum to dryness to furnish a yellow residue that is dissolved in diethyl ether (750 mL). This solution is washed with water (200 mL), dried (magnesium sulfate, $MgSO_4$), and evaporated to dryness to afford 51.9 g (89.5%) of O-ethyl S-[2-oximino-2-(p-chlorophenyl)ethyl]dithiocarbonate as a yellow amorphous solid.

C.   N-Hydroxy-4-(p-chlorophenyl)thiazole-2(3H)-thione.   O-Ethyl S-[2-oximino-2-(p-chlorophenyl)ethyl]dithiocarbonate (56.0 g, 0.19 mol) is placed in a 500-mL round-bottomed flask that is equipped with a magnetic stir bar. Diethyl ether (120 mL) is added and the slurry is treated at 0°C in small portions with solid anhydrous zinc chloride, $ZnCl_2$, 79.1 g, 0.58 mol) at such a rate that the solvent does not boil constantly (Note 8). After the addition is complete, the flask is stoppered with a drying tube ($CaCl_2$) and stirring is continued for 48 hr at 20°C. The reaction mixture turns into a clear, dark brown solution that solidifies toward the end of the reaction. The flask is immersed in an ice bath and treated dropwise with 5.5 M hydrochloric acid (140 mL, Note 9).   The precipitate dissolves immediately.   Stirring is continued for 30 min at 0°C whereupon a tan-colored solid separates. This material is collected by filtration. It is washed with small portions of diethyl ether (total of 110 mL) and dried to afford 39.8 g (86%) of N-hydroxy-4-(p-chlorophenyl)thiazole-2(3H)-thione (Note 10). The crude material is transferred to a 2-L, round-bottomed flask equipped with a reflux condenser. 2-Propanol (760 mL) is added and the reaction mixture is heated to reflux. Once a clear solution is obtained the heat source is immediately removed (Note 11).   The solution is allowed to cool to room temperature.   Precipitation of N-hydroxy-4-(p-chlorophenyl)thiazole-2(3H)-thione is completed by immersing the flask for 30 min in an acetone-dry ice bath (-78°C).   The product is collected by filtration and dried to afford 21.9 g (53.5%) of N-hydroxy-4-(p-chlorophenyl)thiazole-2(3H)-thione as tan crystals (Notes 12, 13).

2. Notes

1. A three-necked, round-bottomed flask with Claisen head may be used instead. p-Chloroacetophenone (97%), hydroxylamine hydrochloride (purum p.a. ≥98%), potassium O-ethyl xanthogenate (≥98%), bromine (puriss. p.a., ≥99.5%), and hydrobromic acid [puriss. 48% (w/w)] were obtained from Fluka Chemika and used as

received. All solvents [acetic acid (reagent grade, 99-100%, Merck & Company, Inc.), diethyl ether (purum, ≥99%, stabilized with 0.0001% of 2,6-di-tert-butyl-p-cresol, Fluka Chemika), acetone (purum, ≥99%, Fluka Chemika), 2-propanol (purum, ≥99%, Fluka Chemika), and ethanol (purum, 96%, 2% ethyl methyl ketone, and 0.5% isobutyl methyl ketone, Fluka Chemika)] were used without further purification. Zinc chloride was used as received from Riedel de Haen (puriss., 98-100%).

2. The temperature of the reaction mixture is kept at 15°C throughout the reaction by adding portions of crushed ice to the water bath. If the temperature drops below 10°C the solvent starts to solidify; above 25°C products of dibromination are formed.

3. Bromine should be handled in a well-ventilated hood. Addition of bromine usually is complete within 75 min.

4. ω-Bromo-p-chloroacetophenone was used without further drying in the oxime-forming reaction. Drying of this product from step (A) at 20°C and 10^{-3} mbar (7.5 x 10^{-4} mm) affords 58.3 g of ω-bromo-p-chloroacetophenone. ω-Bromoacetophenones are lachrymators. Skin contact should be avoided.

5. A 4-cm stirring bar is adequate.

6. The two reactions in Step A were taken from the literature[2,3] and were improved for the present synthesis. All previously related syntheses of arylacetophenone oximes found in the literature have used a threefold excess of hydroxylamine hydrochloride that is, however, not necessary in the present protocol.

7. Drying ω-bromo-p-chloroacetophenone oxime was carried out in the dark since otherwise this material turns pink in light. Wrapping the flask with aluminum foil provides adequate light protection for this purpose.

8. Addition of $ZnCl_2$ is exothermic. The reaction mixture may bubble upon addition of each portion of $ZnCl_2$, but it should not boil constantly. After addition of $ZnCl_2$ is complete, a 4.8-5 M solution of $ZnCl_2$ is obtained that leads to a maximum yield of the title

compound. It was essential to use diethyl ether as solvent in this step. Also no excess diethyl ether should be added after the addition of $ZnCl_2$ is complete.

9. 5.5 M Hydrochloric acid was prepared from 80 mL of concentrated hydrochloric acid and 60 mL of water. This step should be carried out in a well-ventilated hood since an unpleasant smell develops.

10. To some people, N-hydroxy-4-(p-chlorophenyl)-thiazole-2(3H)-thione has a musty odor. Thus, the submitters recommend that the compound be handled and stored in a ventilated place. All drying operations in Step C were carried out at $20°C/10^{-2}$ mbar (7.5 × 10^{-3} mm). If prolonged storage of the title compound is required, the use of an amber-colored vial is recommended.

11. The solution should not be heated longer than necessary to obtain a clear solution. According to differential thermal analysis, neat N-hydroxy-4-(p-chlorophenyl)thiazole-2(3H)-thione decomposes at $138 \pm 2°C$ without melting.

12. Analytical data for N-hydroxy-4-(p-chlorophenyl)thiazole-2(3H)-thione is as follows: IR (CCl_4) cm^{-1}: 3124, 2942, 1603, 1581, 1556, 1500, 1488, 1404, 1355, 1301, 1214, 1174, 1095, 1062, 1016, 976; UV/Vis (EtOH) λ_{max} (log ε): 309 (4.16), 240 nm (4.20); ^{1}H NMR (250 MHz, $CDCl_3$) δ: 6.68 (s, 1 H), 7.47 (m_c, 2 H), 7.62 (m_c, 2 H), 11.72 (br s, 1 H, OH); ^{13}C NMR (100 MHz, (DMSO-d_6) δ: 106.4, 127.6, 128.7, 130.1, 134.4, 140.3, 179.2; Anal. Calcd for $C_9H_6ClNOS_2$ (243.7): C, 44.35; H, 2.48; N, 5.75; S, 26.31. Found: C, 44.35; H, 2.47; N, 5.74; S, 26.39.

13. The submitters report that a second crop of the title compound (7.7 g, 16%) of similar quality separates from the mother liquor upon concentration of the volume of the solution to 100 mL and storage at −20°C overnight. This was not checked.

Waste Disposal Information

All toxic materials were disposed of in accordance with "Prudent Practices in the Laboratory"; National Academy Press; Washington, DC, 1995.

3. Discussion

N-Hydroxy-4-(p-chlorophenyl)thiazole-2(3H)-thione (**1**) is a valuable starting material for the synthesis of N-alkoxy derivatives **2** by a number of well-elaborated O-alkylation procedures (Figure 1).[4,5]

Figure 1

Ar = p-ClC$_6$H$_4$

R = 4-penten-1-yl

R' = tetrahydrofuryl,

tetrahydropyryl

Y-Z = Bu$_3$Sn-**H**

Cl$_3$C-**Br**

I$_2$HC-**I**

Substituted heterocycles **2** serve as sources of oxygen-centered radicals **3** for mechanistic and synthetic purposes. This method was recently applied as a key step in the synthesis of a muscarine alkaloid.[6] N-Alkoxy-4-(p-chlorophenyl)thiazole-2(3H)-thiones **2** offer significant advantages compared with their best current alternatives, the N-alkoxypyridine-2(1H)-thiones[7,8] since they combine two rare properties required for efficient radical precursors. On the one hand they are sufficiently stable to allow safe handling and storage in standard glassware without being prone to photochemical decomposition or to thermal rearrangement, **2** → **6**. This O-alkyl → S-alkyl shift is a significant decomposition pathway for substituted N-benzylpyridine-2(1H)-thiones.[9] In contrast to their thermal stability, thiazolethiones **2** efficiently liberate oxygen-centered radicals **3** in chain reactions upon photolysis.[4] In this sense, N-alkoxy-4-(p-chlorophenyl)thiazole-2(3H)-thiones are superior to many existing alkoxyl radical precursors.

233

A smaller scale and less efficient synthesis of the phenyl derivative of **1** has been reported.[10] The method described for the title compound **1** provides excellent synthetic access to the p-chlorophenyl-substituted thiazolethione **1**. It offers higher yields in every step of the synthesis, reduces the amount of hydroxylamine hydrochloride used in Step A to a third and avoids the use of halogenated solvents and methanol. Further, only one purification step is necessary at the very end of the synthesis to afford pure thione **1**. The protocol for N-hydroxy-4-(p-chlorophenyl) thiazole-2(3H)-thione has also been applied to syntheses of the respective p-substituted phenyl derivatives **7-9**[4] from the respective acetophenones (Figure 2)

Figure 2

X	Compound	Yield
OCH_3	7	46%
CH_3	8	49%
Cl	1	67%
NO_2	9	42%

1. Institut für Organische Chemie, Universität Würzburg, 97074 Würzburg, Germany.

2. Masaki, M.; Fukui, K.; Ohta, M. *J. Org. Chem.* **1967** *32*, 3564.

3. Amschler, U.; Schultz, O.-E. *Arzneim.-Forsch.* **1972**, *22*, 2095.

4. Hartung, J.; Schwarz, M.; Svoboda, I.; Fuess, H.; Duarte, M. T. *Eur. J. Org. Chem.* **1999**, 1275.

5. Hartung, J. *Eur. J. Org. Chem.* **2001**, 619.

6.   Hartung, J.; Kneuer, R. *Eur. J. Org. Chem.* **2000**, 1677.

7.   Hay, B. P.; Beckwith, A. L. J. *J. Org. Chem.* **1989**, *54*, 4330.

8.   Hartung, J.; Gallou, F. *J. Org. Chem.* **1995**, *60*, 6706.

9.   Hartung, J.; Hiller, M.; Schmidt, P. *Chem. Eur. J.* **1996**, *2*, 1014.

10. Barton, D. H. R.; Crich, D.; Kretzschmar, G. *J. Chem. Soc., Perkin Trans. 1*
     1986, 39.

Appendix

Chemical Abstracts Nomenclature (Collective Index Number);
(Registry Number)

N-Hydroxy-4-(p-chlorophenyl)thiazole-2(3H)-thione: 2(3H)-Thiazolethione, 4-(4-chlorophenyl)-3-hydroxy- (11); (105922-93-8)

ω-Bromo-p-chloroacetophenone oxime: Ethanone, 2-bromo-1-(4-chlorophenyl)- oxime (12); (136978-96-6)

p-Chloroacetophenone: Ethanone, 1-(4-chlorophenyl)- (8,9); (99-91-2)

Hydrobromic acid (8,9); (10035-10-6)

Bromine (8,9); (7726-95-6)

Hydroxylamine hydrochoride (8); Hydroxylamine, hydrochloride (9); (5470-11-1)

O-Ethyl S-[oximino-2-(p-chlorophenyl)ethyl]dithiocarbonate: Carbonodithioic acid, S-[2-(4-chlorophenyl)-2-(hydroximino)ethyl], O-ethyl ester (14); (195213-53-7)

Potassium O-ethyl xanthate or Potassium O-ethyl xanthogenate: Carbonodithioic acid O-ethyl ester, potassium salt (8,9); (140-89-6)

Zinc chloride (8,9); (7646-85-7)

PREPARATION OF 1-BUTYL-3-METHYL IMIDAZOLIUM-BASED ROOM TEMPERATURE IONIC LIQUIDS

[1H-Imidazolium, 1-butyl-3-methyl-, chloride (1−); 1H-Imidazolium, 1-butyl-3-methyl-, tetrafluoroborate (1−); 1H-Imidazolium, 1-butyl-3-methyl-, hexafluorophosphate (1−)]

A.

Me—N⎓N + ⌃⌄CI →(MeCN, 80°C)→ Me—N⊕N—Bu, CI⁻

B.

Me—N⊕N—Bu, CI⁻ + KBF₄ →(H₂O)→ Me—N⊕N—Bu, BF₄⁻

C.

Me—N⊕N—Bu, CI⁻ + KPF₆ →(H₂O)→ Me—N⊕N—Bu, PF₆⁻

Submitted by Jairton Dupont, Crestina S. Consorti, Paulo A. Z. Suarez, and Roberto F. de Souza.[1]

Checked by Susan L. Fulmer, David P. Richardson, Thomas E. Smith, and Steven Wolff.

1. Procedure

Caution! 1-Chlorobutane is an irritant and a possible carcinogen.

A. *1-Butyl-3-methylimidazolium chloride.* A 2-L, three-necked, round-bottomed flask is equipped with a heating oil bath, a nitrogen inlet adapter, an internal thermometer adapter, an overhead mechanical stirrer, and a reflux condenser. The flask is flushed with nitrogen and charged with 151.5 g (1.85 mol, 1 equiv) of freshly distilled N-methylimidazole (Note 1), 100 mL of acetonitrile (CH$_3$CN, Note 2) and 220 g (2.4 mol, 1.3 equiv) of 1-chlorobutane (Note 3), and brought to a gentle reflux (75-80°C internal temperature). The solution is heated under reflux for 48 hr (Note 4) and then cooled to room temperature (Note 5). The volatile material is removed from the resulting yellow solution under reduced pressure (Note 6). The remaining light-yellow oil is re-dissolved in dry acetonitrile (250 mL) and added dropwise via cannula to a well-stirred solution of 1000 mL of dry ethyl acetate (Note 7) and one seed crystal of 1- butyl-3-methylimidazolium chloride (Note 8) placed in a 2-L, three-necked, round-bottomed flask, equipped with a nitrogen inlet adapter and an overhead mechanical stirrer (Note 9). The imidazolium salt begins to crystallize exothermically almost immediately, and after the addition of the acetonitrile solution is completed, the flask is cooled at -30°C for 2 hr. The supernatant solution is removed via filtration through a filter cannula and the resulting white solid (Note 10) is dried under reduced pressure (0.1 mbar, 0.001 mm) at 30°C for 6 hr to afford 1-butyl-3-methylimidazolium chloride 289.5 g (89%), mp 66-67°C (Notes 11 and 12).

B. *1- Butyl-3-methylimidazolium tetrafluoroborate, (BMI · BF$_4$)*. A 1-L, one-necked, round-bottomed flask (Note 13) is charged with 91.6 g (0.52 mol, 1 equiv) of finely powdered 1- butyl-3-methylimidazolium chloride, 66.1 g (0.52 mol, 1 equiv) of potassium tetrafluoroborate (Note 14) in 200 mL of distilled water. The reaction mixture is stirred at room temperature for 2 hr affording a heterogeneous mixture (Note 15). The water is removed under reduced pressure (0.1 bar, 0.1 mm) at 80°C until constant weight. To the remaining suspension are added 100 mL of dichloromethane (Note 16) and 35 g of anhydrous magnesium sulfate. After 1 hr of standing the suspension is filtered and the volatile material is removed under reduced pressure (0.1 bar, 0. 1 mm) at 30°C for 2 hr to afford 107.5 g (0.48 mol, 91%) of 1- butyl-3-methylimidazolium tetrafluoroborate as a light yellow, viscous liquid, mp -74°C (Notes 17 and 18).

C.   1- Butyl-3-methylimidazolium hexafluorophosphate.   A 1-L, one-necked, round-bottomed flask (Note 13) is charged with 65.6 g (0.37 mol, 1 equiv) of 1- butyl-3-methylimidazolium chloride, and 69.3 g (0.37 mol, 1 equiv) of potassium hexafluorophosphate (Note 19) in 70 mL of distilled water. The reaction mixture is stirred at room temperature for 2 hr affording a two-phase system. The organic phase is washed with 3 x 50 mL of water and dried under reduced pressure (0.1 mbar, 0.001 mm). Then 100 mL of dichloromethane and 35 g of anhydrous magnesium sulfate are added. After 1 hr, the suspension is filtered and the volatile material is removed under reduced pressure (0.1 bar, 0. 1 mm) at 30°C for 2 hr to afford 86.4 g (0.29 mol, 81%) of 1- butyl-3-methylimidazolium hexafluorophosphate as a light yellow viscous liquid, mp 10°C (Notes 20 and 21).

2. Notes

1. N-Methylimidazole is commercially available. The submitters used a product from Aldrich Chemical Company, Inc. (99%) dried over potassium hydroxide (KOH) pellets and distilled (210-212°C).

2. Acetonitrile (Merck) is distilled over phosphorus pentoxide (P_2O_5).

3. 1-Chlorobutane (Merck) is used as received.

4. The external temperature should not exceed 85°C.

5. The consumption of N-methylimidazole ($R_f = 0.4$) can be followed by TLC using Merck Silica gel 60 and ethyl acetate as eluent (developed using I_2).

6. The volatile material (acetonitrile and the excess of 1-chlorobutane) is captured in a liquid nitrogen trap. This solution (35 wt% in 1-chlorobutane, determined by GC) can be stored in a dark flask and used for further synthesis.

7. Ethyl acetate (Merck) is distilled over P_2O_5.

8. The seed crystal is obtained by dissolving a sample (~1 g) of the crude imidazolium salt in a minimum amount of acetonitrile (3 mL); this solution is allowed to stand at -30°C overnight. The checkers observed spontaneous crystallization upon removal of volatile materials.

9. A 150-rpm agitation speed is used and the rate of the addition will determine the morphology of the imidazolium salt (from finely divided powder to solid blocks that are difficult to powder).

10. If solid blocks are obtained they should be ground before drying.

11. Differential scanning calorimetry is performed at a heating rate of 2°C/min from 20°C to 100°C. The checkers used a conventional melting point apparatus.

12. The product has the following spectral properties: ^{1}H NMR (300 MHz, CDCl$_3$) δ: 0.80 (t, 3 H, $^3J_{HH}$ = 7.3), 1.23 (m, 2 H), 1.75 (m, 2 H), 3.98 (s, 3 H), 4.19 (t, 2 H, $^3J_{HH}$ = 7.4), 7.46 (s, 1 H), 7.63 (s, 1 H), 9.55 (s, 1 H); ^{13}C NMR (75 MHz, CDCl$_3$) δ: 13.6, 19.6, 32.3, 36.6, 49.8, 122.3, 124.0, 137.8; IR (neat film/NaCl plate) cm^{-1}: 3137, 3046, 2959, 2935, 2873, 1571, 1465, 1382, 1336, 1172.

13. The reaction is performed in air without any special precaution.

14. Potassium tetrafluoroborate (Strem Chemicals Inc.) was used as received. Sodium tetrafluoroborate can be also used although it is much more expensive.

15. Although BMI·BF$_4$ is soluble in water at room temperature, the presence of potassium chloride (KCl) gives a salting-out effect, affording two phases. By adding more water a homogeneous colorless solution can be obtained.

16. Dichloromethane (Merck and Co., Inc.) was used as received.

17. A glass transition (-74°C) is obtained by differential scanning calorimetry performed at the cooling rate of 10°C/min from 20°C to -150°C followed by an isothermal at this temperature for 10 min and then heated to 30°C, at the same heating rate. When the cooling rate was decreased to 2 or 1°C/min, the crystallization transition at -73°C was barely observable.

18. The product has the following spectral properties: ^{1}H NMR (300 MHz, acetone-d_6) δ: 0.95 (t, 3 H, $^3J_{HH}$ = 7.3), 1.37 (m, 2 H), 1.93 (m, 2 H), 4.07 (s, 3 H), 4.40 (t, 2 H, $^3J_{HH}$ = 7.1), 7.79 (s, 1 H), 7.85 (s, 1 H), 9.55 (s, 1 H); ^{13}C NMR (75 MHz, acetone-d_6) δ: 13.1, 19.3, 32.2, 35.9, 49.4, 122.7, 124.0, 138.9; IR (neat film/NaCl plate) cm^{-1}: 3160, 3119, 2963, 2938, 2876, 1573, 1171, 1059. For comparison with the literature data see Ref. 2.

19. Potassium hexafluorophosphate (Strem Chemicals Inc.) is used as received. Sodium hexafluorophosphate can be also used although it is much more expensive.

20. A glass transition (-75°C) and two broad bands close to 0°C are obtained by differential scanning calorimetry performed at the cooling rate of 10°C/min from 20°C to -150°C followed by an isothermal at this temperature for 10 min and then heated to 30°C at the same rate. If the heating rate is lowered to 2 or 1°C/min, a crystallization peak is obtained at 10°C.

21. The product has the following spectral properties: ^{1}H NMR (300 MHz, acetone-d_6) δ: 0.96 (t, 3 H, $^3J_{HH}$ = 7.3), 1.37 (m, 2 H), 1.93 (m, 2 H), 4.05 (s, 3 H), 4.36 (t, 2 H, $^3J_{HH}$ = 7.3), 7.68 (s, 1 H), 7.74 (s, 1 H), 8.95 (s, 1 H); ^{13}C NMR (75 MHz, acetone-d_6) δ: 13.0, 19.3, 32.1, 36.0, 49.6, 122.7, 124.1, 137.0; IR (neat film/NaCl plate) cm^{-1}: 3171, 3125, 2965, 2939, 2878, 1571, 1167, 836. For comparison with the literature data see Ref. 2.

Waste Disposal Information

All toxic materials were disposed of in accordance with "Prudent Practices in the Laboratory"; National Academy Press; Washington, DC, 1995.

3. Discussion

The primary advantage in the first step of the method described here (using 1-chlorobutane diluted in MeCN) is that it eliminates long reaction periods[3] and allows the use of secondary alkyl halides without competitive elimination reactions. For example, the reaction of sec-butyl bromide with N-methylimidazole using the classical method (in neat alkyl halide) produces, along with the desired product, 20-30% of butenes and 1-methylimidazole hydrobromide. In the second step, the use of water as solvent allows the anion metathesis reaction to be quantitative in a very short time and allows the easy purification of the ionic liquids. Moreover, employing the potassium salt avoids the use of corrosive and difficult to handle hexafluorophosphoric acid[3] and the expensive silver tetrafluoroborate.[4]

The ionic liquids 1-butyl-3-methylimidazolium tetrafluoroborate (BMI·BF$_4$) and 1-butyl-3-methylimidazolium hexafluorophosphate (BMI·PF$_6$) have a broad application as "green" solvents for organic synthesis,[5] extraction technologies,[6] electrochemistry,[7] biphasic organometallic catalysis[8] and as stationary phases for GC.[9] In particular these room temperature ionic liquids are highly thermal- and electrochemically stable, they possess negligible vapor pressure, have relatively low viscosity and high density (see Table).[10] The most important advantage of the use of these ionic liquids as solvents, in particular for biphasic organometallics, is that it allows the facile separation of the products from the reaction (in most of the cases by simple decanting) and the recovered ionic catalyst solution can be reused. Moreover, ionic liquids can improve or promote reactions that occur with difficulty or do not occur at all in classical organic solvents.

TABLE[10]

PHYSICAL CHEMICAL PROPERTIES FOR 1-BUTYL-3-METHYLIMIDAZOLIUM BASED IONIC LIQUIDS

X	mp (°C)	η_{30} (P)	ρ_{30} (g mL^{-1})	K_{60} (S cm^{-1})10^{-2}	EW (V)
BF$_4$	-74	2.33	1.15	0.864	6.1
PF$_6$	10	3.12	1.37	0.656	7.0

η_{30} = viscosity at 30°C; ρ_{30} = density at 30°C; K_{60} = electrical conductivity at 60°C; EW = electrochemical window.

1. Laboratory of Molecular Catalysis, Instituto de Química, UFRGS, Av. Bento Goncalves, 9500 Porto Alegre 91501-970 RS Brazil.

2. Suarez, P. A. Z.; Dullius, J. E. L.; Einloft, S.; de Souza, R. F.; Dupont, J. *Polyhedron* **1996**, *15*, 1217.

3. Gordon, C. M.; Holbrey, J. D.; Kennedy, A. R.; Seddon, K. R. *J. Mater. Chem.* **1998**, *8*, 2627.

4. Wilkes, J. S.; Zaworotko, M. J. *J. Chem. Soc., Chem. Commun.* **1992**, 965.

5. Earle, M. J.; McCormac, P. B.; Seddon, K. R. *J. Chem. Soc., Chem. Commun.* **1998**, 2245.

6. (a) Blanchard, L. A.; Hancu, D.; Beckman, E. J.; Brennecke, J. F. *Nature*, **1999**, *399*, 28; (b) Huddleston, J. G.; Willauer, H. D.; Swatloski, R. P.; Visser, A. E.; Rogers, R. D. *J. Chem. Soc., Chem. Commun.* **1998**, 1765.

7. McEwen, A. B.; Ngo, E. L.; LeCompte, K.; Goldman, J. L. *J. Electrochem. Soc.* **1999**, *146*, 1687.

8. (a) Welton, T. *Chem. Rev.* **1999**, *99*, 2071; (b) Adams, C. J.; Earle, M. J.; Seddon, K. R. *J. Chem. Soc., Chem. Commun.* **1999**, 1043; (c) Dyson, P. J.; Ellis, D. J.; Parker, D. G.; Welton, T. *J. Chem. Soc., Chem. Commun.* **1999**, 25; (d) Chen, W.; Xu, L.; Chatterton, C.; Xiao, J. *J. Chem. Soc., Chem. Commun.* **1999**, 1247.

9. Armstrong, D. W.; He, L.; Liu, Y.-S. *Anal. Chem.* **1999**, *71*, 3873.

10. Dullius, J. E. L.; Suarez, P. A. Z.; Einloft, S.; de Souza, R. F.; Dupont, J.; Fischer, J.; De Cian, A. *Organometallics* **1998**, *17*, 815.

Appendix
Chemical Abstracts Nomenclature (Collective Index Number); (Registry Number)

1- Butyl-3-methylimidazolium chloride: 1H-Imidazolium, 1-butyl-3-methyl-, chloride (10); (79917-90-1)

1- Butyl-3-methylimidazolium tetrafluoroborate: 1H-Imidazolium, 1-butyl-3-methyl-, tetrafluoroborate(1-) (13); (174501-65-6)

1- Butyl-3-methylimidazolium hexafluorophosphate: 1H-Imidazolium, 1-butyl-3-methyl-, hexafluorophosphate(1-) (13); (174501-64-5)

N-Methylimidazole: Imidazole, 1-methyl- (8); 1H-Imidazole, 1-methyl (9); (616-47-7)

1-Chlorobutane: Butane, 1-chloro- (8,9); (109-69-3)

Potassium tetrafluoroborate: Borate(1-), tetrafluoro-, potassium (8,9); (14075-53-7)

Potassium hexafluorophosphate: Phosphate(1-), hexafluoro-, potassium (8,9); (17084-13-8)

4-METHOXYCARBONYL-2-METHYL-1,3-OXAZOLE

(4-Oxazolecarboxylic acid, 2-methyl-, methyl ester)

A.

$$\underset{NH_2Cl}{\overset{OMe}{\diagdown}} \quad + \quad CINH_3 \diagup CO_2Me \quad \xrightarrow[76\%]{Et_3N, \ CH_2Cl_2} \quad \underset{N}{\overset{OMe}{\diagdown}} CO_2Me$$

1

B.

$$\underset{N}{\overset{OMe}{\diagdown}} CO_2Me \quad + \quad HCO_2Me \quad \xrightarrow[THF, \ ether]{t\text{-}BuOK} \quad \underset{N}{\overset{OMe}{\diagdown}} \overset{OK}{\underset{CO_2Me}{\diagup}}$$

1   **2**

C.

$$\underset{N}{\overset{OMe}{\diagdown}} \overset{OK}{\underset{CO_2Me}{\diagup}} \quad \xrightarrow[38\%]{AcOH} \quad \underset{N}{\overset{O}{\diagup}} CO_2Me$$

2   **3**

Submitted by James D. White, Christian L. Kranemann, and Punlop Kuntiyong.[1]

Checked by Mitsuru Kitamura and Koichi Narasaka.

1. Procedure

A. Methyl α-[(methoxyethylidene)amino]acetate (**1**). A flame-dried, 500-mL, two-necked, round-bottomed flask is equipped with a stir bar, rubber septum, and an argon inlet. The flask is charged with methyl acetimidate hydrochloride (10.0 g, 91 mmol) (Note 1) and dry dichloromethane (140 mL) (Note 2). The stirred suspension is cooled to 0°C and solid methyl glycinate hydrochloride (11.5 g, 91 mmol, Note 1) is added in one portion with a powder funnel under a stream of Ar. After the mixture is stirred for 45 min at 0°C, a solution of dry triethylamine (12.7 mL, 91 mmol) (Note 2) in dry dichloromethane (11 mL) is added via syringe pump during 2.5 hr. Stirring is continued for 5 hr while the mixture is

244

allowed to warm slowly to room temperature (Note 3). Water (30 mL, pH 7 buffered) is added, giving a clear biphasic mixture (Note 4). The phases are separated in a 250-mL separatory funnel, and the aqueous phase is extracted with dichloromethane (2 x 15 mL). The combined organic phases are washed with pH 7 buffered water (1 x 17 mL) and brine (1 x 17 mL). After the organic solution is dried over anhydrous magnesium sulfate, it is filtered and concentrated under reduced pressure, leaving 11.50 g of the crude product as a colorless solid. Distillation of this material (41 mm, 135°C) gives 10.10 g (76%) of pure methyl α-[(methoxyethylidene)amino]acetate (**1**, Note 5).

  B.   Potassium methyl α-[(methoxyethylidene)amino]-β-hydroxyacrylate (**2**).   A flame-dried, 2-L, three-necked, round-bottomed flask is equipped with a stir bar, rubber septum, and an argon inlet. The flask is charged with a solution of potassium tert-butoxide (7.81 g, 70 mmol) (Note 1) in dry tetrahydrofuran (THF, 200 mL) (Note 2), and the solution is stirred at -10°C for 15 min. A solution of methyl α-[(methoxyethylidene)amino]acetate (10.10 g, 70 mmol) and methyl formate (5.0 mL, 84 mmol) (Note 1) in dry THF (50 mL) is added via a syringe pump during 20 min. After a further 5 min at -10°C, dry diethyl ether (750 mL) (Note 2) is added via a cannula, resulting in the formation of a yellowish precipitate. Stirring is continued for 2 hr at 0°C and the cold solution is filtered through a Schlenk tube (Note 6) under argon. The pale yellow filter cake is washed under argon with dry diethyl ether (3 x 40 mL), and the cake is dried under an argon stream and then under reduced pressure. The solid is transferred from the Schlenk tube to a wide mouthed vessel under an argon atmosphere. The resultant crude potassium methyl α-[(methoxyethylidene)amino-β-hydroxyacrylate (**2**) is used directly for the next step (Note 7).

  C.   4-Methoxycarbonyl-2-methyl-1,3-oxazole (**3**).   A flame-dried, 50-mL, two-necked, round-bottomed flask is equipped with a stir bar, reflux condenser, rubber septum, and an argon inlet. The flask is charged with glacial acetic acid (15 mL) which is heated to reflux.   To this is added crude potassium methyl α-[(methoxyethylidene)-amino]-β-hydroxyacrylate, prepared above, in one portion with a powder funnel. Material which adheres to the wall of the funnel and the flask is washed into the mixture with a stream of acetic anhydride. The mixture is stirred at reflux for 1.5 hr, then allowed to cool and carefully poured into a 250-mL Erlenmeyer flask containing a saturated aqueous solution of sodium bicarbonate (50 mL) (Note 7). The pH of the solution is adjusted to 8 by further addition of solid sodium bicarbonate (Note 8). The solution is extracted with dichloromethane (4 x 30 mL), and the organic extract is dried over anhydrous sodium sulfate, filtered and concentrated under reduced pressure to give 4.48 g of crude product. This is purified b y distillation (41 mm, 150°C) to afford 3.74 g (38% from 1; Note 9) of 4-methoxycarbonyl-2-methyl-1,3-oxazole (**3**, Note 10).

2. Notes

1. Methyl acetimidate hydrochloride, methyl glycinate hydrochloride, potassium tert-butoxide, and methyl formate were purchased from Aldrich Chemical Company, Inc., and were used without further purification. The checkers purchased methyl glycinate hydrochloride from Tokyo Chemical Industry Co. and potassium tert-butoxide and methyl formate from Kanto Chemical Co. Step A is very sensitive to moisture. Ethyl acetimidate hydrochloride is very hygroscopic. It must be dried before use in a desiccator over phosphorus pentoxide (P_2O_5) under reduced pressure and handled under argon.

2. Dichloromethane and triethylamine were freshly distilled from calcium hydride under argon before use. THF and diethyl ether were distilled from sodium and benzophenone under argon. The checkers used THF and diethyl ether as received from Kanto Chemical Co. (reagent grade, <0.005% water).

3. After 2 hr the ice-bath is no longer refilled with fresh ice, allowing the mixture to warm to room temperature during the remaining 3 hr.

4. Stirring for 2 to 3 min is required for complete dissolution of all precipitate.

5. The product is characterized by NMR spectroscopy: ^{1}H NMR (400 MHz, CDCl$_3$) δ: 1.86 (s, 3 H), 3.66 (s, 3 H), 3.71 (s, 3 H), 4.03 (s, 2 H); ^{13}C NMR (100 MHz, CDCl$_3$) δ: 14.5, 50.6, 51.4, 52.2, 164.8, 171.1.

6. Compound **2** is very hygroscopic. ^{1}H NMR (500 MHz, DMSO-d$_6$) δ: 1.60 (s, 3 H), 3.36 (s, 3 H), 3.49 (s, 3 H), 8.57 (s, 1 H).

7. *Caution*: a large amount of carbon dioxide is liberated!

8. Water (50 mL) is added to keep all inorganic salts dissolved.

9. The submitters found that the yield of **3** can be increased to 58% if crude **2**, obtained in step B by rotary evaporation of the solvent (rather than filtration through a Schlenk tube followed by washing with ether), is taken directly into hot glacial acetic acid in step C. This procedure minimizes exposure of hygroscopic **2** to moisture. These changes were not checked.

10. The product is characterized by NMR-spectroscopy: ^{1}H NMR (400 MHz, (CDCl$_3$) δ: 2.31 (s, 3 H), 3.70 (s, 3 H), 7.97 (s, 1 H); ^{13}C NMR (100 MHz, CDCl$_3$) δ: 13.5, 51.8, 133.0, 143.6, 161.4, 162.2.

3. Discussion

The method described for the preparation of 4-methoxycarbonyl-2-methyl-1,3-oxazole is that of Cornforth,[2] and is widely applicable to the synthesis of 2-substituted 1,3-oxazole-4-carboxylates.[3] The appropriate imidate hydrochloride required for step A is obtained from the reaction of a nitrile with an alcohol in the presence of hydrochloric acid (eq. 1).[4] A different synthesis of 2-substituted 1,3-oxazole-4-carboxylates employing rhodium-catalyzed heterocycloaddition of a diazomalonate to a nitrile has been described in *Organic Syntheses* by Helquist,[5] but appears to be less general than the present route.

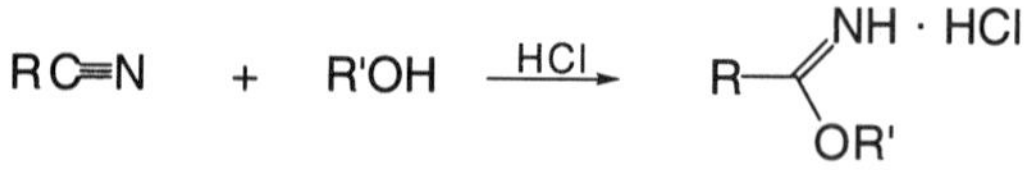

$$RC{\equiv}N \quad + \quad R'OH \quad \xrightarrow{\text{HCl}} \quad R \overset{NH \cdot HCl}{\underset{OR'}{\diagdown}} \qquad (1)$$

New methods for the synthesis of 2,4-disubstitued oxazoles are summarized in a recent review.[6] 2-Alkyl-1,3-oxazoles bearing alkyl, aryl, or acyl substitution at C4 are common substructures in natural products.[7] Examples include macrolides such as rhizoxin (4),[8] hennoxazole A (5),[9] and phorboxazole A (6),[10] as well as many cyclic peptides that incorporate an oxazole subunit presumably derived from serine.[11]

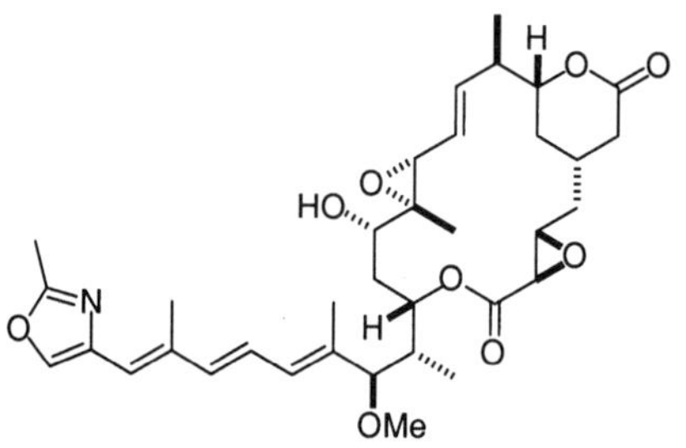

Rhizoxin (4)

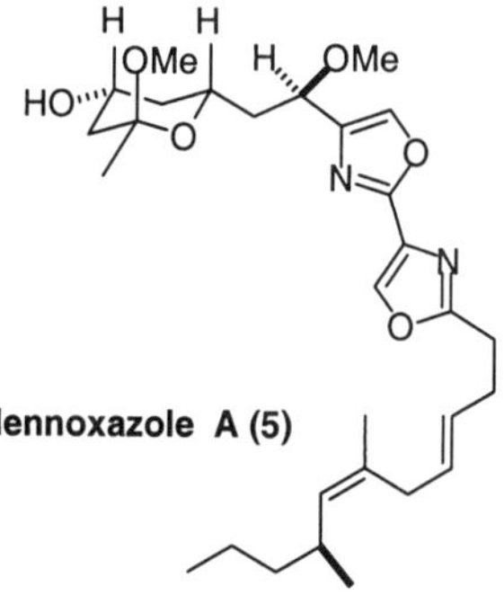

Hennoxazole A (5)

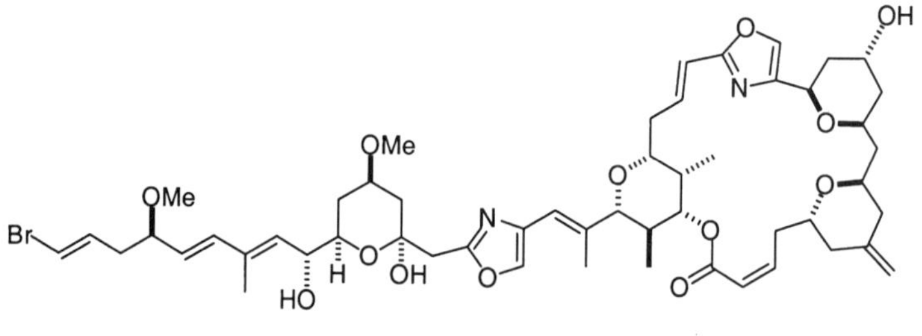

Phorboxazole A (6)

4-Methoxycarbonyl-2-methyl-1,3-oxazole (3) is metalated exclusively at C5 with n-butyllithium.[3] Selective functionalization at the methyl group of 3 can be achieved with N-bromosuccinimide to yield the 2-bromomethyl derivative 8. The latter affords a route to

2,4-disubstituted oxazoles that are not immediately accessible through the Cornforth synthesis. Thus, **8** undergoes displacement with sodium phenylsulfinate to give sulfone **9**, which can then be transformed to aldehyde **10**.

1. Department of Chemistry, Oregon State University, Corvallis, OR 97331-4003.

2. Cornforth, J. W.; Cornforth, R. H. *J. Chem. Soc.* **1947**, 96.

3. Meyers, A. I.; Lawson, J. P.; Walker, D. G.; Lindermann, R. J. *J. Org. Chem.* **1986**, *51*, 5111.

4. (a) Molina, P.; Lopez-Leonardo, C.; Llamas-Botia, J.; Foces-Foces, C.; Llamas-Sais, A. L. *Synthesis* **1995**, 449; (b) Law, H.; Dukat, M.; Teitler, M.; Lee, D. K. H.; Mazzocco, L.; Kamboj, R.; Rampersad, V.; Prisinzano, T.; Glennon, R. A. *J. Med. Chem.* **1998**, *41*, 2243.

5. Tullis, J. S.; Helquist, P. *Org. Synth., Coll. Vol. IX* **1998**, 155.

6. Gilchrist, T. L. *J. Chem. Soc., Perkin Trans. 1* **1998**, 615. A comprehensive listing of citations to oxazole chemistry, including synthesis, is contained in reference 5.

7. (a) Lewis, J. R. *Nat. Prod. Rep.* **1998**, *15*, 371, 417; (b) Roy, R. S.; Gehring, A. M.; Milne, J. C.; Belshaw, P. J.; Walsh, C. T. *Nat. Prod. Rep.* **1999**, *16*, 249.

8. Iwasaki, S.; Namikoshi, M.; Kobayashi, H.; Furukawa, J.; Okuda, S. *Chem. Pharm. Bull.* **1986**, *34*, 1387.

9. Ichiba, T.; Yoshida, W. Y.; Scheuer, P. J.; Higa, T.; Gravalos, D. G. *J. Am. Chem. Soc.* **1991**, *113*, 3173.

10. Searle, P. A.; Molinski, T. F. *J. Am. Chem. Soc.* **1995**, *117*, 8126.

11. Lewis, J. R. *Nat. Prod. Rep.* **1999**, *16*, 389.

Appendix

Chemical Abstracts Nomenclature (Collective Index Number);

(Registry Number)

4-Methoxycarbonyl-2-methyl-1,3-oxazole: 4-Oxazolecarboxylic acid, 2-methyl-, methyl ester (11); (85806-67-3)

Methyl α-[(methoxyethylidene)amino]acetate: Glycine, N-(1-methoxyethylidene)-, methyl ester (10); (64991-38-4)

Methyl acetimidate hydrochloride: Acetimidic acid, methyl ester, hydrochloride (8); Ethanimidic acid, methyl ester, hydrochloride (9); (14777-27-6)

Methyl glycinate hydrochloride: ALDRICH: Glycine methyl ester hydrochloride: Glycine, methyl ester, hydrochloride (8,9); (5680-79-5)

Triethylamine (8); Ethanamine, N,N-diethyl- (9); (121-44-8)

Potassium methyl α-[(methoxyethylidene)amino]-β-hydroxyacrylate: Propanoic acid, 2-[(1-methoxyethylidene)amino]-3-oxo-, methyl ester, ion(1-), potassium (11); (105205-36-5)

Potassium tert-butoxide: tert-Butyl alcohol, potassium salt (8); 2-Propanol, 2-methyl-, potassium salt (9); (865-47-4)

Methyl formate: Formic acid, methyl ester (8,9); (107-31-3)

Acetic acid (8,9); (64-19-7)

Acetic anhydride (8); Acetic acid,anhydride (9); (108-24-7)

[4+3] CYCLOADDITION IN WATER.

SYNTHESIS OF 2,4-endo,endo-DIMETHYL-8-OXABICYCLO[3.2.1]OCT-6-EN-3-ONE

[8-Oxabicyclo[3.2.1]oct-6-en-3-one, 2,4-dimethyl-, (endo,endo)-]

A. (structure **1**) + SO_2Cl_2 $\xrightarrow[45°C]{CCl_4}$ (structure **2**)

B. (structure **2**) + (furan) $\xrightarrow[H_2O, \text{ rt}]{Et_3N}$ (structure **3**)

Submitted by Mark Lautens and Giliane Bouchain.[1]

Checked by Patrick Foyle and Steven Wolff.

1. Procedure

A. 2-Chloropentan-3-one, **2**. A 500-mL, two-necked, round-bottomed flask, containing a magnetic stirring bar, is equipped with a 100-mL pressure-equalizing addition funnel and a reflux condenser fitted with a calcium chloride trap. The flask is charged with 85 mL (69.40 g, 0.80 mol) of pentan-3-one **1** (Note 1) and 200 mL of carbon tetrachloride and the mixture is heated to 45°C in an oil bath. The addition funnel is charged with 71 mL (0.88 mol, 1.1 mol equiv) of sulfuryl chloride that is added dropwise over a period of 2 hr. The resulting mixture is stirred for 3 hr at 45°C, carbon tetrachloride is removed by distillation under atmospheric pressure at 85°C (Note 2) and the residue is purified by distillation under reduced pressure (Note 3). After a forerun at 65-80°C (62 mm) containing mainly 3-

pentanone, 2-chloropentan-3-one **2** is collected (bp 80-102°C, 62 mm) as a pale yellow liquid (77.0 g, 80%) (Notes 4, 5).

   B.   *2,4-endo,endo-Dimethyl-8-oxabicyclo[3.2.1]oct-6-en-3-one,* **3**. A 500-mL round-bottomed flask is equipped with a magnetic stirring bar and a 50-mL pressure-equalizing addition funnel. The flask is charged with 15.0 g (0.12 mol) of **2** (Note 6), 36.1 mL (0.50 mol, 4 mol equiv) of furan and 125 mL of distilled water. The mixture is stirred vigorously at room temperature and triethylamine (18.05 mL, 0.13 mol, 1.05 mol equiv) is placed in the addition funnel and added dropwise to the reaction over a period of 30 min (Note 7). The reaction mixture is stirred for 12 hr and quenched by adding a saturated solution of ammonium chloride (50 mL). The mixture is poured into a 500-mL separatory funnel. The layers are separated and the aqueous layer is extracted with dichloromethane (3 x 50 mL). The combined organic layers are washed with brine, dried over anhydrous magnesium sulfate, and the organic layer is filtered and concentrated under reduced pressure. The crude reaction mixture is resubjected to the reaction conditions by adding additional furan and triethylamine (1.05 mol equiv based on the amount of unreacted starting material as determined by ^{1}H NMR), (Notes 8, 9). After 5 hr of vigorous stirring at room temperature, the reaction is quenched by adding a saturated solution of ammonium chloride (50 mL). The layers are separated, the aqueous layer is extracted with dichloromethane (3 x 25 mL) and the combined organic layers are washed with brine and dried over anhydrous magnesium sulfate. After filtration, the solvent is removed with a rotatory evaporator and the crude reaction mixture is dried for 2 hr on a high vacuum pump. The oil is cooled to -20°C in the freezer overnight and pale yellow crystals are filtered on a Buchner funnel and washed with cold pentane. The mother liquors are concentrated, placed on a high vacuum pump for 5 hr, and cooled in the freezer. The procedure is repeated once more so that a total of 6.10 g (33%) of pure **3** (Note 10) is obtained.

   The filtrate is concentrated under reduced pressure and the residue is purified by flash column chromatography on silica gel [500 mL, 230-400 mesh, eluted sequentially with hexane/ethyl acetate (90% to 70%)] to afford the cycloadduct **3** as a white powder (2.25 g, 12%); total 8.35 g (45%) (Note 11).

2. Notes

1. Pentan-3-one, furan and sulfuryl chloride were purchased from Aldrich Chemical Company, Inc., and used without further purification.

2. The condenser and the addition funnel are removed and replaced by a simple distillation head and a condenser.

3. The residue is cooled to room temperature and transferred into a 200-mL round-bottomed flask equipped with a 6"-Vigreux column, distillation head, condenser and a cow receiver to be purified by distillation under reduced pressure.

4. A small quantity of 2,4-dichloropentan-3-one (5%) is obtained with the α-monochloropentanone **2** during the distillation. The NMR analysis for 2-chloropentan-3-one **2** is described by Wyman and Kaufman[2].

5. The monochloroketone must be stored in a refrigerator.

6. Trace amounts of dichloroketone do not interfere with the [4+3] cycloaddition.

7. The mixture becomes a pale orange, biphasic solution. This mixture becomes red if the addition of triethylamine is too fast.

8. The reaction cannot be monitored by TLC because the monochloro ketone cannot be visualized.

9. The starting material remaining is usually about 40% of the crude weight.

10. Spectral data for the α,α-cycloadduct **3** are identical to that reported by Hoffmann and coworkers[3].

11. The submitters report the isolation of 0.75 g (4%) of the corresponding exo-exo isomer; the checkers could not isolate this material in pure form.

Waste Disposal Information

All toxic materials were disposed of in accordance with "Prudent Practices in the Laboratory"; National Academic Press; Washington, DC, 1995.

3. Discussion

Seven membered ring heterocycles are valuable synthetic intermediates for various natural products. The uses and synthesis of [4+3] cycloadducts have been the subject of extensive reviews.[4] The usual method to prepare these compounds is the [4+3] cycloaddition between an oxyallylic cation and various dienes. The procedure described here provides a simple and efficient method for the construction of 8-oxabicyclo[3.2.1]oct-6-ene on a large scale in water using common reagents and mild conditions. To date, the two main routes reported in the literature involve the generation of oxyallylic cations starting from α,α'-dihalo ketones[5] or α-monohalo ketones.[6] All these reactions were carried out using sophisticated promotors [Zn-B(OEt)$_3$,[7] CeCl$_3$-SnCl$_2$,[8] etc.] or an expensive and highly toxic reductive agent [Fe$_2$(CO)$_9$],[9] under nitrogen in anhydrous solvents. Lubineau and Bouchain[10] reported the [4+3] cycloaddition in water using α,α'-dibromo ketones or α-monochloro ketones to provide [3.2.1]oxabicyclic compounds in good yields. Moreover, these conditions afford a good to excellent stereoselectivity in favor of the α,α-cis isomers. By using the monochloro ketone in excess (2 equiv), the yield of the [4+3] cycloadduct is 88% based on furan, but this procedure, which uses furan in excess, is less cost effective. In order to optimize the yield, after a cycloaddition reaction, the crude mixture containing some starting monochloropentan-3-one was worked-up and resubjected to the same conditions without any intermediate purification. Moreover the pure α,α-cycloadduct is easily obtained by crystallization.

In conclusion, this is the most practical and least expensive method available for the synthesis of [3.2.1]oxabicyclic compounds.

1. University of Toronto, Davenport Research Laboratories, Toronto, ON, M5S, 3H6, Canada.

2. Wyman, D. P.; Kaufman, P. R. *J. Org. Chem.* **1964**, *29*, 1956).

3. Hoffmann, H. M. R. *Angew. Chem., Int. Ed. Engl.* **1973**, *12*, 819-835; Hoffmann, H. M. R. *Angew. Chem., Int. Ed. Engl.* **1984**, *23*, 1-19; Noyori, R.; Hayakawa, Y. *Org. React.* **1983**, *29*, 163-344; Mann, J. *Tetrahedron* **1986**, *42*, 4611-4659; Chiu, P.; Lautens, M.; *Top. Curr. Chem.* **1997**, *190*, 1.

4. Rawson, D. I.; Carpenter, B. K.; Hoffmann, H. M. R. *J. Am. Chem. Soc.* **1978**, *100*, 1765

5. Fort, A. W. *J. Am. Chem. Soc.* **1962**, *84*, 4979-4981; Cookson, R. C.; Nye, M. J. *Proc. Chem. Soc.* **1963**, 129-130.

6. Föhlisch, B.; Gehrlach, E.; Herter, R. *Angew. Chem., Int. Ed. Engl.* **1982**, *21*, 137; Föhlisch, B.; Gehrlach, E.; Stezowski, J. J.; Kollat, P.; Martin, E.; Gottstein, W. *Chem. Ber.* **1986**, *119*, 1661-1682.

7. Hoffmann, H. M. R.; Iqbal, M. N. *Tetrahedron Lett.* **1975**, 4487-4490.

8. Fukuzawa, S.; Fukushima, M.; Fujinami, T.; Sakai, S. *Bull. Chem Soc.. Jpn.* **1989**, *62*, 2348-2352.

9. Takaya, H.; Makino, S.; Hayakawa, Y.; Noyori, R. *J. Am. Chem. Soc.* **1978**, *100*, 1765-1777.

10. Lubineau, A.; Bouchain, G, *Tetrahedron Lett.* **1997**, *38*, 8031-8032, and citations therein.

Appendix

Chemical Abstracts Nomenclature (Collective Index Number); (Registry Number)

2,4-endo,endo-Dimethyl-8-oxabicyclo[3.2.1]oct-6-en-3-one: 8-Oxabicyclo[3.2.1]oct-6-en-3-one, 2,4-dimethyl-, (endo,endo)- (9); (37081-58-6)

2-Chloropentan-3-one: 3-Pentanone, 2-chloro- (8,9); (17042-21-6)

Carbon tetrachloride: CANCER SUSPECT AGENT (8); Methane, tetrachloro- (9); (56-23-5)

Sulfuryl chloride (8,9); (7791-25-5)

Furan (8,9); (110-00-9)

Triethylamine (8); Ethanamine, N,N-diethyl- (9); (121-44-8)

Unchecked Procedures

Accepted for checking during the period September 1, 2000 through September 1, 2001. An asterisk(*) indicates that the procedure has been subsequently checked.

Previously, *Organic Syntheses* has supplied these procedures upon request. However, because of the potential liability associated with procedures which have not been tested, we shall continue to list such procedures but requests for them should be directed to the submitters listed.

2883   2-(Hydroxymethyl)-2-cyclohexen-1-one.
F. Rezgui and M. M. El Gaïed, Département de Chimie, Faculté des Sciences, 1060 Campus Universitaire, Tunis, Tunisie.

2884R   Poly(ethylene glycol) Bis-(6-(methylsulfinyl)hexanoate).
J. M. Harris, Y. Liu, and J. C. Vederas, Department of Chemistry, University of Alberta, Edmonton, Alberta, T6G 2G2, Canada.

2894   Preparation of O-Allyl-N-(9)-anthracenylmethylcinchonidinium Bromide as a Phase Transfer Catalyst for the Enantioselective Alkylation of Glycine Benzophenone Imine tert-Butyl Ester.
E. J. Corey and M. C. Noe, Department of Chemistry, Harvard University, 12 Oxford Street, Cambridge, MA 01238.

2894A   Enantioselective Alkylation of Glycine Benzophenone Imine tert-Butyl Ester.
E. J. Corey and M. C. Noe, Department of Chemistry, Harvard University, 12 Oxford Street, Cambridge, MA 01238.

2935R   Synthesis of N-(tert-Butoxycarbonyl)-ß-Iodoalanine Methyl Ester: A Useful Building Block in the Synthesis of Nonnatural α-Amino Acids via Palladium Catalyzed Cross Coupling Reactions.
R. F. W. Jackson and M. Perez-Gonzalez, Department of Chemistry, Bedson Building, The University of Newcastle upon Tyne, New Castle upon Tyne, NF1 7RU, UK.

2937   Asymmetric Synthesis of N-tert-Butoxycarbonyl α-Amino Acids: Synthesis of (5S,6R)- and (5R,6S)-4-tert-Butoxycarbonyl-2,3-diphenylmorpholin-6-one.
R. M. Williams, D. E. DeMong, P. J. Sinclair, D. Chen, and D. Zhai, Department of Chemistry, Colorado State University, Fort Collins, CO 80523.

2938   Efficient Asymmetric Synthesis of N-tert-Butoxycarbonyl α-Amino Acids Using (tert-Butoxy)carbonyl-2,3-diphenylmorpholin-6-one: Preparation of (R)- Allylglycine.
R. M. Williams, P. J. Sinclair, and D. E. DeMong, Department of Chemistry, Colorado State University, Fort Collins, CO 80523.

2952   Diethylaminotrimethylsilane Mediated Direct 1,4-Addition of Naked Aldehydes: 3R,7-Dimethyl-2R,S-(3-oxobutyl)-6-octenal.
H. Hagiwara, H. Ono, and T. Hoshi, Graduate School of Science and Technology, Niigata University, 8050, 2-nocho, Ikarashi, 950-2181, Japan.

2954   Ring-Closing Metathesis Synthesis of N-Boc-3-Pyrroline.
M. L. Ferguson, D. J. O'Leary, and R. H. Grubbs, Department of Chemistry, Pomona College, 645 N. College Ave., Claremont, CA 91711-6338.

2957   Synthesis of 5-Bromoisoquinoline and 5-Bromo-8-nitroisoquinoline.
W. Hansen and A. H. Gouliaev, NeuroSearch A/S, Pederstrupvej 98, 2750 Ballerup, Denmark.

2959   Generation and [2+2] Cycloadditions of Thio-Substituted Ketenes: trans-1-(4-Methoxyphenyl)-4-phenyl-3-(phenylthio)azetidin-2-one.
R. L. Danheiser, I. Okamoto, M. D. Lawlor, and T. W. Lee, Department of Chemistry, MIT, Room 18-297, Cambridge, MA 02139-4307.

2960   Generation of Nonracemic 2-(t-Butyldimethylsilyloxy)-3-butynyllithium from (S)-Ethyl Lactate: (S)-4-(t-Butyldimethylsilyloxy)-2-pentynol.
J. A. Marshall, M. M. Yanik, and N. D. Adams, Department of Chemistry, The University of Virginia, Charlottesville, VA 22901.

2962   Preparation of 9,10-Dimethoxyphenanthrene and 3,6-Diacetyl-9,10-dimethoxyphenanthrene.
K. Paruch, L. Vyklicky, and T. J. Katz, Chemistry Department, Columbia University, New York, NY 10027.

2963   Helicenebisquinones: Synthesis of a [7]Helicenebisquinone.
K. Paruch, L. Vyklicky, and T. J. Katz, Chemistry Department, Columbia University, Hanemeyer Hall, Mail Code 3112, New York, NY 10027.

2964   n-Butyl 2,2-Difluorocyclopropanecarboxylate. Preparation and Use of a New Difluorocarbene Reagent.
W. R. Dolbier, Jr., F. Tian, J.-X. Duan, and Q.-Y. Chen, Department of Chemistry, University of Florida, P.O. Box 117200, Gainesville, FL 32611-7200.

2965   (S)-4-(1-Methylethyl)-5,5-diphenyloxazolidin-2-one.
M. Brenner, L. La Vecchia, T. Leutert, D. Seebach, Laboratorium für Organische Chemie, ETH, Universitätstr. 16, CH-8092 Zürich, Switzerland.

2967   Generation of an Acetylene-Titanium Alkoxide Complex, Preparation of (Z)-1,2-Dideuterio-1-(trimethylsilyl)-1-hexene.
H. Urabe, D. Suzuki and F. Sato, Department of Biomolecular Engineering, Tokyo Institute of Technology, 4259, Nagatsuta-cho, Midori-ku, Yokohama, Kanagawa, 226-8501, Japan.

2968   (R)-(+)-3,4-Dimethylcyclohex-2-en-1-one.
J. D. White and U. M. Grether, Department of Chemistry, Oregon State University, 153 Gilbert Hall, Corvallis, OR 97331-4003.

2971   2-Chloro-1, 3-bis(dimethylamino) trimethinium Hexafluorophosphate.
Ian W. Davies, Jean-Francois Marcoux, and Jeremy Taylor, Process Research, R800-B263, Merck & Co., Inc., P.O. Box 2000, Rahway, NJ 07065.

2972   Intra- and Intermolecular Kulinkovich Cyclopropanation Reactions of Carboxylic Esters with Olefins.
Se-Ho Kim, Moo Je Sung, and Jin Kun Cha, Department of Chemistry, University of Alabama, Loyd Hall, Box 870336, Tuscaloosa, AL 35487-0336.

2974   1-Hydroxy-3-Phenyl-2-Propanone.

Marjorie S. Waters, Kelley Snelgrove, and Peter Maligres, Department of Process Research, Merck Research Laboratories, Merck & Co., Inc., Rahay, NJ 07065.

CUMULATIVE AUTHOR INDEX
FOR VOLUMES 75, 76, 77, 78, AND 79

This index comprises the names of contributors to Volume **75, 76, 77, 78,** and **79** only. For authors to previous volumes, see either indices in Collective Volumes I through IX or the single volume entitled *Organic Syntheses, Collective Volumes I-VIII, Cumulative Indices*, edited by J. P. Freeman.

Kuntiyong, P., **79**, 244
Kurosawa, W., **79**, 186

Lajoie, G. A., **79**, 216
Lamba, J. J. S., **78**, 51
Lang, M., **78**, 113
Lang-Fugmann, S., **78**, 113
Larchevêque, M., **75**, 37
Larrow, J. F., **75**, 1; **76**, 46
Larsen, R. D., **78**, 36
Laurent, A., **76**, 159
Lautens, M., **79**, 251
La Vecchia, L., **76**, 12
Lebel, H., **76**, 86
Le Goffic, F., **76**, 123
Levin, M. D., **77**, 249
Ley, S. V., **75**, 170; **77**, 212
Liebeskind, L. S., **77**, 135
Linder, M.R., **79**, 154
Lipshutz, B. H., **76**, 252
Liu, H., **76**, 189
Liu, Y.-Z., **76**, 151
Love, J. C., **78**, 51
Luo, F.-T., **75**, 146
Luo, Y., **79**, 216
Lynam, N., **77**, 220
Lynch, K. M., **75**, 89, 98
Lynch, S. M., **78**, 202

MacKinnon, J., **75**, 124
Mack, R. A., **77**, 45
Macor, J., **77**, 45
Makowski, T. W., **78**, 63
Mangeney, P., **76**, 23
Mann, J., **75**, 139

Marshall, J. A., **76**, 263; **77**, 98; **79**, 59
Martinelli, M. J., **75**, 223
Maryanoff, B. E., **75**, 215
Maryanoff, C. A., **75**, 215
Maumy, M., **76**, 133
Mazerolles, P., **76**, 221
McComsey, D. F., **75**, 215
McDonald, F. E., **79**, 27
McWilliams, J. C., **79**, 43
Meffre, P., **76**, 123
Mellinger, M., **77**, 198
Michl, J., **77**, 249
Mitchell, D., **75**, 53
Miyaura, N., **77**, 176
Modi, D. P., **79**, 165
Moore, H. W., **76**, 189
Moore, J. R., **77**, 12
Mori, A., **75**, 210
Mullins, J. J., **77**, 141
Murata, M., **77**, 176
Myers. A. G., **76**, 57, 178; **77**, 22, 29

Nakai, T., **76**, 151
Nakajima, A., **76**, 199
Nayyar, N. K., **75**, 223
Nelson, J. D., **79**, 165
Neyer, G., **76**, 294
Novak, B. M., **75**, 61
Nowick, J. S., **78**, 220
Noyori, R., **79**, 139

Oevers, S. R., **79**, 196
Ogasawara, M., **79**, 84
Oh, J., **78**, 212
Oh, T., **76**, 101
Ohara, S., **79**, 176

Öhrlein, R., **77**, 236

Oka, H., **79**, 139

Okabe, M., **76**, 275

Olah, G. A., **76**, 294

Osborn, H. M. I., **75**, 170; **77**, 212

O'Sullivan, T., **77**, 220

Padwa, A., **78**, 202

Page, P. C. B., **76**, 37

Panek, J. S., **75**, 78

Park, M., **77**, 153, 162

Paquette, L. A., **75**, 106; **77**, 107

Parlitz, R., **79**, 196

Payack, J. F., **76**, 6; **79**, 19

Peña-Cabrera, E., **77**, 135

Perchet, R. N., **75**, 124

Perrone, D., **77**, 64, 78

Peterson, B. C., **75**, 223

Petit , Y., **75**, 37

Phillips, B. W., **75**, 19

Pierce, M. E., **77**, 12

Podlech, J., **79**, 154

Posakony, J., **76**, 287

Powell, N. A. **78**, 220

Powers, J. P., **77**, 1

Priepke, H. W. M., **75**, 170

Prudhomme, D. R., **77**, 153, 162

Qian, C.-P., **76**, 151

Qun, L., **79**, 11

Ragan, J. A. , **78**, 63

Ravikumar, V. T., **76**, 271

Rawal, V. H., **78**, 152, 160

Reddy, G. V., **77**, 50

Reed, D. P., **78**, 73

Reetz, M. T., **76**, 110

Reid, C., **79**, 216

Reider, P. J., **76**, 1, 6, 46

Rigby, J. H., **77**, 121

Rizzo, C. J., **77**153, 162

Robbins, M. A., **76**, 101

Roberts, E., **76**, 46

Robin, S., **77**, 206

Ronsheim, M. D., **79**, 146

Rose, N. G. W., **79**, 216

Ross, B., **76**, 271

Rousseau, G., **77**, 206

Rousselet, G., **76**, 133

Ruel, F. S., **75**, 69

Ryan, K. M., **76**, 46

Rychnovsky, S. D., **77**, 1

Sadhukhan, S. K., **79**, 52

Saluzzo, C., **77**, 91

Sampognaro, A. J., **77**, 45

Sasai, H., **78**, 14

Sattler, A., **76**, 159

Savage, S. A., **78**, 51

Schick, H., **75**, 116

Schwab, W., **77**, 236

Schwardt, O., **78**, 123

Schwarz, M., **79**, 228

Schwickardi, R., **76**, 110

Seebach, D., **76**, 12

Seed, A. J., **79**, 204

Sehon, C. A., **76**, 263

Seidel, G., **76**, 142

Selva, M., **76**, 169

Senanayake, C. H., **76**, 46

Sengupta, S., **79**, 52

Shahlai, K., **75**, 201

CUMULATIVE SUBJECT INDEX
FOR VOLUMES 75, 76, 77, 78, AND 79

This index comprises subject matter for Volumes **75**, **76**, **77**, **78**, and **79**. For subjects in previous volumes, see either the indices in Collective Volumes I through IX or the single volume entitled *Organic Syntheses, Collective Volumes I-VIII, Cumulative Indices*, edited by J. P. Freeman.

The index lists the names of compounds in two forms. The first is the name used commonly in procedures. The second is the systematic name according to **Chemical Abstracts** nomenclature, accompanied by its registry number in parentheses. Also included are general terms for classes of compounds, types of reactions, special apparatus, and unfamiliar methods.

Most chemicals used in the procedure will appear in the index as written in the text. There generally will be entries for all starting materials, reagents, intermediates, important by-products, and final products. Entries in capital letters indicate compounds, reactions, or methods appearing in the title of the preparation.

ABNORMAL REACTIONS, SUPPRESSION OF, **76**, 228

Acetaldehyde; (75-07-0), **75**, 106

Acetaldehyde dimethyl acetal: Ethane, 1,1-dimethoxy-; (534-15-6), **75**, 46

Acetic acid, glacial; (64-19-7), **75**, 2, 225; **77**, 142

Acetic anhydride: Acetic acid anhydride; (108-24-7), **76**, 70; **77**, 45, 142

Acetone: 2-Propanone; (67-64-1), **76**, 13

Acetonitrile: TOXIC: (75-05-8), **75**, 146; **76**, 24, 47, 67, 191; **77**, 121; **78**, 83

Acetonylacetone: 2,5-Hexanedione (110-13-4), **78**, 64

7α-ACETOXY-(1Hβ, 6Hβ)-BICYCLO[4.4.1]UNDECA-2,4,8-TRIENE:
BICYCLO[4.4.1]UNDECA-3,7,9-TRIEN-2-OL, ACETATE, endo- (±)-;
(129000-83-5), **77**, 121

(E)-1-Acetoxy-1,3-butadiene: 1,3-Butadien-1-ol, acetate, (E)-;
(35694-20-3), **77**, 122

1-Acetoxy-3-(methoxymethoxy)butane: 1-Butanol, 3-(methoxymethoxy)-, acetate;
(167563-42-0), **75**, 177

Acetylacetaldehyde dimethyl acetal: 2-Butanone, 4,4-dimethoxy-; (5436-21-5),
78, 152

Acetylacetonatobis(ethylene)rhodium(I): Rhodium, bis(ethylene)(2,4-
pentanedionato)- (8); Rhodium, bis(η^2-ethene)(2,4-pentanedionato-O,O')-
(9); (12082-47-2), **79**, 84

Acetylacetone: 2,4-Pentanedione; (123-54-6) **78**, 135

ACETYLALLENE: see 3,4-PENTA-1,2-DIEN-2-ONE

9-Acetylanthracene: Ethanone, 1-(9-anthracenyl)-; (784-04-3), **76**, 276

3-Acetyl-6-butoxy-2H-pyran-2,4(3H)-dione: 2H-Pyran-2,4(3H)-dione,
6-butoxydihydro-3-(1-hydroxyethylidene)-; (182616-30-4), **77**, 115

Acetyl chloride; (75-36-5), **75**, 177; **77**, 65, 114; **78**, 99

1-Acetyl-1-cyclohexene: Ethanone, 1-(1-cyclohexen-1-yl)-; (932-66-1), **76**, 203

Acetylenes, terminal, cyanation of, **75**, 148

Acetyl Meldrum's acid: See 5-(1-hydroxyethylidene)-1,3-dioxane-4,6-dione

2-(4'-ACETYLPHENYL)THIOPHENE: ETHANONE, 1-[4-(2-THIENYL)PHENYL]-;
(35294-37-2), **77**, 135

Acyloxyboron intermediates in amide formation, **79**, 180

Acrolein: 2-Propenal; (107-02-8), **77**, 237

Agar plate preparation, **76**, 79

Airfree® reaction flask, **79**, 28, 29

Aliquat 336: Ammonium, methyltrioctyl-, chloride; 1-Octanaminium, N-methyl-N,N-dioctyl-, chloride; (5137-55-3), **76**, 40

Alkyl aryl thioethers, table, **79**, 50

Alkyllithium solutions, to titrate, **76**, 68

Alkynol cycloisomerization, **79**, 27

Allene: 1,2-Propadiene; (463-49-0), **75**, 129

Allenes, stereodefined synthesis of, **76**, 185

Allenylindium reagents, *in situ* formation, enantioenriched, **79**, 59
additions to achiral aldehydes, table, **79**, 67

Allenylzinc reagents (transient), addition to achiral aldehydes, table, **79**, 69

ALLYLATION, CATALYTIC ASYMMETRIC, OF ALDEHYDES, **75**, 12
table, **75**, 17

Allyl bromide: 1-Propene, 3-bromo-; (106-95-6), **76**, 60, 221; **79**, 11

Allyl chloride: 1-Propene, 3-chloro-; (107-05-1), **77**, 254

L-ALLYLGLYCINE: 4-PENTENOIC ACID, 2-AMINO-, (R)-; (54594-06-8), **76**, 57

Allylic alcohols, enantioselective cyclopropanation, **76**, 97

ALLYLINDATION, **77**, 107

Allylmagnesium bromide: Magnesium, bromo-2-propenyl-; (1730-25-2), **76**, 221

Allylmagnesium bromide (ethereal complex solution), **76**, 221

Allylmagnesium bromide (THF complex solution), **76**, 222

Allyltributylstannane: Stannane, tributyl-2-propenyl-; (24850-33-7), **75**, 12

Aluminum chloride; (7446-70-0), **77**, 2; **79**, 204

Amide formation from carboxylic acids and amines, catalyzed by (3,4,5-trifluorophenyl)-boronic acid, **79**, 176
table, **79**, 184
mechanism, **79**, 182

Aminals, **76**, 30

α-Amino aldehydes:
Boc-protected, **76**, 117
N,N-Dibenzyl-protected, **76**, 115

α-Aminocarboxylic acids, **75**, 25; **76**, 57, 123

2-Amino-5-chlorobenzophenone: Methanone, (2-amino-5-chlorophenyl)phenyl-; (719-59-5), **76**, 142

PHOSPHINE, [1,1'-BINAPHTHALENE]-2,2'-DIYLBIS[DIPHENYL-, (R)-; (76189-55-4); **79**, 94 (S)-; (76189-56-5), **76**, 6

rac-2,2'-Bis(diphenylphosphino)-1,1'-binaphthyl- :rac BINAP: Phosphine, [1,1'-binaphthalene]-2,2'-diylbis{diphenyl-; (98327-87-8), **78**, 23; **79**, 43

[1,4-Bis(diphenylphosphino)butane (dppb): Phosphine, 1,4-butanediylbis [diphenyl-; (7688-25-7), **78**, 3

[1,2-Bis(diphenylphosphino)ethane]nickel(II) chloride: Nickel, dichloro[ethanediylbis[diphenylphosphine]-P,P']-, (SP-4-2)-; (14647-23-5), **76**, 7

1,1'-Bis(diphenylphosphino)ferrocene (dppf): Phosphine, 1,1'-ferrocenediylbis[diphenyl- (8); Ferrocene, 1,1'-bis(diphenylphosphino)- (9); (12150-46-8), **79**, 59
[1,1'-Bis(diphenylphosphino)ferrocene]dichloropalladium: Palladium, [1,1'-bis(diphenylphosphino)ferrocene-P,P']dichloro- (10); (72287-26-4), **79**, 59

[1,3-Bis(diphenylphosphino)propane]nickel(II) chloride : Nickel, dichloro[1,3-propanediylbis(diphenylphosphine)-P,P']-;(15629-92-2), **77**, 154; **78**, 43

(R)-(+)-2,2'-Bis(di-p-tolylphosphino)-1,1'-binaphthyl [(R)-Tol-BINAP]: Phosphine, [1,1'-binaphthalene]-2,2'-diylbis[bis(4-methylphenyl)-, (R)- (11); (99646-28-3), **79**, 43

Bis(η-divinyltetramethyldisiloxane)tri-tert-butylphosphineplatinum(0): Platinum, [1,3-bis(η^2-ethenyl)-1,1,3,3-tetramethyldisiloxane][tris(1,1-dimethylethyl)phosphine]-; (104602-18-8), **75**, 79
preparation, **75**, 81

1,2:5,6-Bis-O-(1-methylethylidene)-, O-phenyl carbonothioate-)-α-D-glucofuranose: α-D-Glucofuranose, 1,2:5,6-Bis-O-(1-methylethylidene)-, O-phenyl thiocarbonate; (19189-62-9), **78**, 239

BIS(PINACOLATO)DIBORON: 2,2'-BI-1,3,2-DIOXABOROLANE, 4,4,4',4',5,5,5',5'-OCTAMETHYL-; (73183-34-3), **77**, 176

Bis(ruthenium dichloride-S-BINAP)-triethylamine catalyst: Ruthenium, bis[[1,1'binaphthalene]-2,2'diylbis[diphenylphosphine]-P,P']di-μ-chlorodichloro(N,N-diethylethanamine)di-; (114717-51-0), **77**, 3

Bis(tributyltin) oxide: Distannoxane, hexabutyl-; (56-35-9), **78**, 240

Bis(trichloromethyl) carbonate: See Triphosgene, **75**, 46

(R)-2,2'-Bis(trifluoromethanesulfonyloxy)-1,1'-binaphthyl: Methanesulfonic acid, trifluoro-, [1,1'-binaphthalene]-2.2'-diyl ester, (R)-; (126613-06-7), **78**, 23

(2R-cis)-2-[[1-[3,5-Bis(trifluoromethyl)phenyl]ethenyl]oxy]-3-(4-fluorophenyl)-4-benzylmorpholine: Morpholine, 2-[[1-[3,5-bis(trifluoromethyl)phenyl]-ethenyl]oxy]-3-(4-fluorophenyl)-4-(phenylmethyl)-, (2R,3S)- (9);

3-(4-BROMOBENZOYL)PROPANOIC ACID: BENZENEBUTANOIC ACID, 4-BROMO-γ-OXO- (9); (6340-79-0), **79**, 204

Bromobis(dimethylamino)borane: Boranediamine, 1-bromo-N,N,N',N'-tetramethyl-; (6990-27-8), **77**, 177

1-Bromobutane: Butane, 1-bromo-; (109-65-9), **76**, 87

ω-Bromo-p-chloroacetophenone oxime: Ethanone, 2-bromo-1-(4-chlorophenyl)-, oxime, (12); (136978-96-6), **79**, 228

1-Bromo-5-chloropentane: Pentane, 1-bromo-5-chloro-; (54512-75-3), **76**, 222

1-Bromo-3-chloropropane: Propane, 1-bromo-3-chloro-; (109-70-6), **76**, 222

1-Bromo-1-cyclopropylcyclopropane: 1,1'-Bicyclopropyl, 1-bromo-; (60629-95-0), **78**, 143

Bromofluorination of alkenes, **76**, 159

1-BROMO-2-FLUORO-2-PHENYLPROPANE: BENZENE, (2-BROMO-1-FLUORO-1-METHYLETHYL)-; (59974-27-5), **76**, 159

Bromoform: Methane, tribromo- (8,9); (75-25-2), **75**, 98

6-Bromo-1-hexene: 1-Hexene, 6-bromo-; (2695-47-8), **76**, 223

(S)-(-)-2-Bromo-3-hydroxypropanoic acid: Propanoic acid, 2-bromo-3-hydroxy-, (S)-; (70671-46-4), **75**, 37

Bromomalononitrile: Propanedinitrile, bromo-; (1885-22-9), **75**, 210

preparation, **75**, 211

4-Bromo-3-methylanisole: Benzene, 1-bromo-4-methoxy-2-methyl-; (27060-75-9), **78**, 23

4-Bromo-2-methyl-1-butene: 1-Butene, 4-bromo-2-methyl-; (20038-12-4), **78**, 203

2-Bromonaphthalene: Naphthalene, 2-bromo-; (580-13-2), **76**, 13

1-Bromo-2-naphthoic acid: 2-Naphthoic acid, 1-bromo- (9); (20717-79-7), **79**, 72

8-Bromo-1-octene: 1-Octene, 8-bromo; (2695-48-9), **76**, 224

2-Bromo-1-octen-3-ol: 1-Octen-3-ol, 1-bromo-, (E)-; (52418-90-3), **76**, 263

4-[(4-Bromophenyl)azo]morpholine: Morpholine, 4-[(4-bromophenyl)azo]- (14); (188289-57-8), **79**, 52

1-Bromo-3-phenylpropane: Benzene, (3-bromopropyl)-; (637-59-2), **75**, 155

(Z/E)-1-BROMO-1-PROPENE: 1-PROPENE, 1-BROMO-; (590-14-7), **76**, 214

4-(2-BROMO-2-PROPENYL)-4-METHYL-γ-BUTYROLACTONE: 2(3H)-FURANONE, 5-(2-BROMO-2-PROPENYL)DIHYDRO-5-METHYL-;

(138416-14-5), **75**, 129

2-Bromopyridine: Pyridine, 2-bromo-; (109-04-6), **78**, 53

N-Bromosuccinimide: 2,5-Pyrrolidinedione, 1-bromo-; (128-08-5), **77**, 107; **78**, 234

Bromotrichloromethane: Methane, bromotrichloro; (75-62-7), **75**, 125

1-Bromo-3,4,5-trifluorobenzene: Benzene, 5-bromo-1,2,3-trifluoro- (13); (138526-69-9), **79**, 176

Bromotris(2-perfluorohexylethyl)tin: Stannane,bromotris(3,3,4,4,5,5,6,6,7,7,8,8,8-tridecafluorooctyl)- (13); (175354-31-1), **79**, 1

Butane; (106-97-8), **77**, 231

2,3-Butanedione; (431-03-8), **77**, 249

1-Butanethiol (8,9); (109-79-5), **79**, 43

1-Butanol; (71-36-3), **76**, 48; **78**, 240

2-Butanol; (78-92-2), **76**, 68

N-Boc-L-ALLYLGLYCINE: 4-PENTENOIC ACID, 2-[[(1,1-DIMETHYLETHOXYCARBONYL]AMINO]-, (R)-; (170899-08-8), **76**, 57

(S)-2-[(4S)-N-tert-BUTOXYCARBONYL-2,2-DIMETHYL-1,3-OXAZOLIDINYL]-2-tert-BUTYLDIMETHYLSILOXYETHANAL: 3-OXAZOLIDINECARBOXYLIC ACID, (4-[[(1,1-DIMETHYLETHYL)DIMETHYLSILYL]OXY-2-OXOETHYL]-2,2-DIMETHYL-,1,1-DIMETHYLETHYL ESTER, [S-(R*,R*)]-; (168326-01-0), **77** , 78

(S)-2-{[(4S)-N-tert-Butoxycarbonyl-2,2-dimethyl-1,3-oxazolidinyl]-tert-butyldimethylsiloxy}-1,3-thiazole: 3-Oxazolidinecarboxylic acid, 4-[[[(1,1-dimethylethyl)dimethylsilyl]oxy]-2-thiazolylmethyl]2,2-dimethyl-, 1,1-dimethylethyl ester, [S-(R*,R*)]-; (168326-00-9), **77**, 80

(S)-2-{[(4S)-N-tert-Butoxycarbonyl-2,2-dimethyl-1,3-oxazolidin-4-yl]hydroxymethyl}-1,3-thiazole: 3-Oxazolidinecarboxylic acid, 4-(hydroxy-2-thiazolylmethyl)-2,2-dimethyl-, 1,1-dimethylethyl ester, [(S-(R*,R*)]-; (115822-48-5), **77**, 79

tert-Butyl alcohol: 2-Propanol, 2-methyl-; (75-65-0), **75**, 225; **78**, 114, 203

Butylboronic acid: Boronic acid, butyl-; (4426-47-5), **76**, 87

tert-Butyl chloride: Propane, 2-chloro-2-methyl-; (507-20-0), **75**, 107

n-BUTYL 4-CHLOROPHENYL SULFIDE, **79**, 43

(5S)-(5-O-tert-BUTYLDIMETHYLSILOXYMETHYL)FURAN-2(5H)-ONE: 2(5H)-FURANONE, 5-[[[(1,1-DIMETHYLETHYL)DIMETHYLSILYL]OXY] METHYL]-, (S)-; (105122-15-4), **75**, 140

(1R*,6S*,7S*)-4-tert-BUTYLDIMETHYLSILOXY)-6-(TRIMETHYLSILYL)
BICYCLO[5.4.0]UNDEC-4-EN-2-ONE, **76**, 199

(tert-BUTYLDIMETHYLSILYL)ALLENE: SILANE, (1,1-DIMETHYLETHYL)
DIMETHYL-1,2-PROPADIENYL-; (176545-76-9), **76**, 178

tert-Butyldimethylsilyl chloride: Silane, chloro(1,1-dimethylethyl)dimethyl-;
(18162-48-6), **75**, 140; **76**, 179, 201

1-(tert-Butyldimethylsilyl)-1-(1-ethoxyethoxy)-1,2-propadiene: Silane, (1,1-
dimethylethyl)-dimethyl-1-(1-ethoxyethoxy)-1,2-propadienyl-;
(86486-46-6), **76**, 201

1-(tert-Butyldimethylsilyl)-1-(1-ethoxyethoxy)-3-trimethylsilyl-
1,2-propadiene (crude), **76**, 180

3-(tert-Butyldimethylsilyl)-2-propyn-1-ol: 2-Propyn-1-ol, 3-[(1,1-dimethylethyl)-
dimethylsilyl]-; (120789-51-7), **76**, 179

tert-Butyldimethylsilyl trifluoromethanesulfonate: Methanesulfonic acid,
trifluoro-, (1,1-dimethylethyl)dimethylsilyl ester; (69739-34-0), **77**, 80

(E)-1-(tert-Butyldimethylsilyl)-3-trimethylsilyl-2-propen-1-one: Silane,
(1,1-dimethylethyl)dimethyl[1-oxo-3-(trimethylsilyl)-2-propenyl]-, (E)-;
(83578-66-9), **76**, 202; (Z)-, **76**, 207

(2R,3S,4S)-1-(tert-Butyldiphenylsilyloxy)-2,4-dimethyl-5-hexyn-3-ol: 5-Hexyn-3-ol,
1-[[(1,1-dimethylethyl)diphenylsilyl]oxy]-2,4-dimethyl-, (2R,3S,4S)- (14);
(220634-80-0), **79**, 59

(R)-3-(tert-Butyldiphenylsilyloxy)-2-methylpropanal: Propanal, 3-[[(1,1-
dimethylethyl)diphenylsilyl]oxy]-2-methyl-, (2R)- (12);
(112897-04-8), **79**, 59

2-BUTYL-6-ETHENYL-5-METHOXY-1,4-BENZOQUINONE:
2,5-CYCLOHEXADIENE-1,4-DIONE, 5-BUTYL-3-ETHENYL-
2-METHOXY-; (134863-12-0), **76**, 189

Butyllithium: Lithium, butyl-; (109-72-8), **75**, 22, 116, 201; **76**, 37, 59, 151, 179,
191, 201, 202, 203, 214, 230, 239; **77**, 23, 31, 32, 98, 231; **78**, 15, 82,
114; **79**, 11, 35, 84

Butylmagnesium bromide: Magnesium, bromobutyl-; (693-03-8), **76**, 87

tert-Butyllithium: Lithium, (1,1-dimethylethyl)-; (594-19-4), **77**, 212; **78**, 53; **79**, 11

assay, **79**, 13

tert-Butyl N-(3-methyl-3-butenyl)-N-(2-furyl)carbamate: Carbamic acid, 2-furanyl
(3-methyl-3-butenyl)-, 1,1-dimethylethyl ester (212560-95-7), **78**, 204

tert-Butyl methyl ether: Ether, tert-butyl methyl; Propane, 2-methoxy-2-
methyl-; (1634-04-4), **76**, 276; **77**, 199

283

2-Chlorophenol: Phenol, 2-chloro; (95-57-8), **76**, 271

4-Chlorophenol: Phenol, 4-chloro-; (106-48-9), **78**, 27; **79**, 43

2-CHLOROPHENYL PHOSPHORODICHLORIDOTHIOATE: PHOSPHORODICHLORIDOTHIOIC ACID, O-(2-CHLOROPHENYL) ESTER; (68591-34-4), **76**, 271

4-Chlorophenyl trifluoromethanesulfonate: Methanesulfonic acid, trifluoro-, 4-chlorophenyl ester; (29540-84-9), **78**, 27; **79**, 43

Chloroplatinic acid hexahydrate: Platinate (2-), hexachloro-, dihydrogen, (OC-6-11)-; (16941-12-1), **75**, 81

3-Chloropropionitrile: Propanenitrile, 3-chloro-; (542-76-7), **77**, 186

6-Chloro-1-pyrrolidinocyclohexene: Pyrrolidine, 1-(6-chloro-1-cyclohexen-1-yl)-; (35307-20-1), **78**, 212

N-Chlorosuccinimide: 2,5-Pyrrolidinedione, 1-chloro-; (128-09-6), **76**, 124

Chlorotrimethylsilane: Silane, chlorotrimethyl-; (75-77-4), **75**, 146; **76**, 24, 252; **77**, 26, 91,163, 199; **78**, 83, 99, 105

Chlorotriphenylmethane: Benzene, 1,1',1"-(chloromethylidyne)tris-; (76-83-5), **75**, 184

Chromium hexacarbonyl: Chromium carbonyl (OC-6-11)-; (13007-92-6), **77**, 121

trans-Cinnamaldehyde: 2-Propenal, 3-phenyl-, (E)-; (14371-10-9), **76**, 214

Cinnamyl alcohol: 2-propen-1-ol, 3-phenyl-; (104-54-1), **76**, 89

Citric acid monohydrate: 1, 2, 3-Propanetricarboxylic acid, 2-hydroxy-monohydrate; (5949-29-1), **75**, 31

Clathrate, **76**, 14

Cobalt nitrate hexahydrate: Nitric acid, cobalt (2+ salt), hexahydrate; (10026-22-9), **76**, 79

2,4,6-Collidine: Pyridine, 2,4,6-trimethyl-; (108-75-8), **77**, 207

Copper; (7440-50-8), **76**, 94, 255

Copper(I) bromide-dimethyl sulfide complex: Copper, bromo[thiobis[methane (9); (54678-23-8), **79**, 11

COPPER-CATALYZED CONJUGATE ADDITION OF ORGANOZINC REAGENTS TO α,β-UNSATURATED KETONES, **76**, 252

Copper(I) chloride: Copper chloride; (7758-89-6), **76**, 133; **78**, 64

Copper(II) chloride: Copper chloride; (7447-39-4), **77**, 213

Copper(II) chloride dihydrate: Copper chloride, dihydrate; (10125-13-0), **77**, 81

78, 113

Diisopropylamine: 2-Propanamine, N-(1-methylethyl)-; (108-18-9), **75**, 116; **76**, 59, 203, 239; **77**, 23, 31, 32, 98; **78**, 82

Diisopropyl ether: Propane, 2,2'-oxybis-; (108-20-3), **75**, 130

N,N-Diisopropylethylamine: 2-Propanamine, N-ethyl-N-(1-methylethyl)-; (7087-68-5), **75**, 177; **77**, 68, 100; **79**, 216

1,2:5,6-Di-O-isopropylidene-D-glucose: α-D-Glucofuranose, 1,2:5,6-bis-O-(1-methylethylidene)-; (582-52-5), **78**, 239

Dilithium tetrachloromanganate: Manganate (2–), tetrachloro-, dilithium, (I-4)-; (57384-24-4), **76**, 240

(+)-[(8,8-Dimethoxycamphoryl)sulfonyl]imine: 3H-3a,6-Methano-2,1-benzisothiazole,4,5,6,7-tetrahydro-7,7-dimethoxy-8,8-dimethyl-, 2,2-dioxide, [3aS]-; (131863-80-4), **76**, 40

(+)-[(8,8-Dimethoxycamphoryl)sulfonyl]oxaziridine: 4H-4a,7-Methanoxazirino[3,2-i][2,1]benzisothiazole, tetrahydro-8,8-dimethoxy-9,9-dimethyl-, 3,3-dioxide, [2R-(2α,4aα,7α,8aS*)]-; (131863-82-6), **76**, 38, 39

Dimethoxyethane: Ethane, 1,2-dimethoxy-; (110-71-4), **76**, 89

(3,4-Dimethoxyphenyl)acetonitrile: Benzeneacetonitrile, 3,4-dimethoxy-; (93-17-4), **76**, 133

2-(3,4-Dimethoxyphenyl)-N,N-dimethylacetamidine; (240797-77-7), **76**, 133

2,2-Dimethoxypropane: Propane, 2,2-dimethoxy-; (77-76-9), **77**, 66

N,N-Dimethylacetamide, **79**, 72

Dimethylamine: Methanamine, N-methyl-; (124-40-3), **76**, 93, 133; **77**, 176; **78**, 152

(E)-4-Dimethylamino-3-buten-2-one: 3-Buten-2-one, 4-(dimethylamino)-, (E)-; (2802-08-6), **78**, 152

1-DIMETHYLAMINO-3-tert-BUTYLDIMETHYLSILOXY-1,3-BUTADIENE: 1,3-BUTADIEN-1-AMINE, 3-[[(1,1-DIMETHYLETHYL)DIMETHYLSILYL]OXY]-N,N-DIMETHYL-, (E)-; (194233-66-4), **78**, 152, 161

(2S)-(-)-3-exo-(DIMETHYLAMINO)ISOBORNEOL: [(2S)-(-)-DAIB]: BICYCLO[2.2.1]HEPTAN-2-OL, 3-(DIMETHYLAMINO)-1,7,7-TRIMETHYL-, [1R-(exo, exo]- (11); (103729-96-0), **79**, 130
enantiomeric purity determination, **79**, 135
catalyst for enantioselective addition of diethylzinc to aldehydes, **79**, 142
table, **79**, 143
recovery, **79**, 141

1-(3-Dimethylamino)propyl-3-ethylcarbodiimide hydrochloride: 1,3-Propanediamine, N'-(ethylcarbonimidoyl)-N,N-dimethyl-, monohydrochloride; (25952-53-8), **77**, 35

4-Dimethylaminopyridine: HIGHLY TOXIC: 4-Pyridinamine, N,N-dimethyl-;
(1122-58-3), **75**, 184; **76**, 70; **77**, 7, 38, 72, 80; **79**, 72

4-DIMETHYLAMINO-N-TRIPHENYLMETHYLPYRIDINIUM CHLORIDE:
PYRIDINIUM, 4-(DIMETHYLAMINO)-1-(TRIPHENYLMETHYL)-,
CHLORIDE; (78646-25-0), **75**, 184

1,3-Dimethyl-6H-benzo[b]naphtho[1,2d]pyran-6-one: 6H-Benzo[b]naphtho[1,2-
d]pyran-6-one, 1,3-dimethyl- (13); (138435-72-0), **79**, 72

4,4'-Dimethyl-2,2'-bipyridine: 2,2'-Bipyridine, 4,4'-dimethyl-; (1134-35-6), **78**, 82

(1R,2R)-(+)-N,N'-Dimethyl-1,2-bis(3-trifluoromethyl)phenyl-
1,2-ethanediamine; (120263-19-6), **76**, 127

Dimethyl carbonate: Carbonic acid, dimethyl ester; (616-38-6), **76**, 170

Dimethyl 3-cyclopentene-1,1-dicarboxylate; 3-Cyclopentene-1,1-dicarboxylic
acid, dimethyl ester; (84646-68-4), **75**, 197

N,N'-DIMETHYL-1,2-DIPHENYLETHYLENEDIAMINE: 1,2-ETHANEDIAMINE,
N,N'-DIMETHYL-1,2-DIPHENYL, (R*,S*)-; (60509-62-8); [R-(R*,R*)-;
(118628-68-5); [S-(R*,R*)]; (70749-06-3), **76**, 23

N-[(1,1-Dimethylethoxy)carbonyl]-N,O-isopropylidene-L-serinol:
3-Oxazolidinecarboxylic acid, 4-(hydroxymethyl)-2,2-dimethyl-,
1,1-dimethylethyl ester, (R)-; (108149-63-9), **77**, 66

N-[(1,1-Dimethylethoxy)carbonyl]-L-serine methyl ester: L-Serine,
N-[(1,1-dimethylethoxy)carbonyl]-, methyl ester; (2766-43-0), **77**, 65

1,1-DIMETHYLETHYL 2,2-DIMETHYL-(S)-4-FORMYLOXAZOLIDINE-
3-CARBOXYLATE: 3-OXAZOLIDINECARBOXYLIC ACID, 4-FORMYL-
2,2-DIMETHYL-, 1,1-DIMETHYLETHYL ESTER, (S)-; (102308-32-7),
77, 64, 79, 89

3-(1,1-Dimethylethyl) 4-methyl (S)-2,2-dimethyl-3,4-oxazolidine-
dicarboxylate: 3,4-Oxazolidinedicarboxylic acid, 2,2-dimethyl-,
3-(1,1-dimethylethyl) 4-methyl ester, (S)-; (108149-60-6), **77**, 66

N,N-Dimethylformamide: CANCER SUSPECT AGENT: Formamide,
N,N-dimethyl- (8,9); (68-12-2), **75**, 162; **77**, 35, 37, 221; **78**, 36, 64

N,N-DIMETHYLHOMOVERATRYLAMINE: BENZENEETHANAMINE,
3,4-DIMETHOXY-N,N-DIMETHYL-; (3490-05-9), **76**, 133

(R,R)-Dimethyl O,O-isopropylidenetartrate: 1,3-Dioxolane-4,5-dicarboxylic
acid, 2,2-dimethyl-, dimethyl ester, (4R-trans)-; (37031-29-1), **76**, 13

Dimethyl malonate: Propanedioic acid, dimethyl ester; (108-59-8), **75**, 195

2,7-DIMETHYLNAPHTHALENE: NAPHTHALENE, 2,7-DIMETHYL-; (582-16-1),
78, 42

Epoxidation, **75**, 153; **76**, 50, 53; **78**, 225
    asymmetric, **75**, 9

EPOXIDATION CATALYST, ENANTIOSELECTIVE, **75**, 1

Epoxides, optically acitive, **75**, 37 **76**, 107; **77**, 1, 91,

Ergosterol: Ergosta-5,7,22-trien-3-ol, (3β)-; (57-87-4), **76**, 276

3-ETHENYL-4-METHOXYCYCLOBUTENE-1,2-DIONE: 3-CYCLOBUTENE-
    1,2-DIONE, 3-ETHENYL-4-METHOXY-; (124022-02-2), **76**, 189

1-(1-Ethoxyethoxy)-1,2-propadiene: 1,2-Propadiene, 1-(1-ethoxyethoxy)-;
    (20524-89-4), **76**, 200

1-(1-Ethoxyethoxy)-1-propyne: 1-Propyne, 3-(1-ethoxyethoxy)-; (18669-04-0),
    76, 200

Ethyl acetate: Acetic acid, ethyl ester; (141-78-6), **77**, 30

Ethyl [2-^{13}C] acetate: Acetic-2-^{13}C acid, ethyl ester; (58735-82-3), **78**, 114

ETHYL (R)-2-AZIDOPROPIONATE: PROPANOIC ACID, 2-AZIDO-, ETHYL
    ESTER, (R)-; (124988-44-9), **75**, 31
    assay of optical purity, **75**, 33

Ethyl benzoate: Benzoic acid, ethyl ester; (93-89-0), **75**, 215

(-)-(S)-Ethyl 2-(benzyloxy)propanoate: Propanoic acid, 2-(phenylmethoxy)-, ethyl
    ester, (S)-; (54783-72-1), **78**, 178

Ethyl bromide: Bromoethane; (74-96-4), **75**, 38

Ethyl 5-bromovalerate: Pentanoic acid, 5-bromo-, ethyl ester; (14660-52-7),
    76, 255

Ethyl chloroformate: Carbonochloridic acid, ethyl ester; (541-41-3), **78**, 114;
    79, 154

ETHYL 5-CHLORO-3-PHENYLINDOLE-2-CARBOXYLATE: 1H-INDOLE-
    2-CARBOXYLIC ACID, 5-CHLORO-3-PHENYL-, ETHYL ESTER;
    (212139-32-2), **76**, 142

Ethyl 3-chloropropionate: Propionic acid, 3-chloro-, ethyl ester (8); Propanoic
    acid, 3-chloro-, ethyl ester (9); (623-71-2), **79**, 35

ETHYL 3-(4-CYANOPHENYL)PROPIONATE: BENZENEPROPANOIC ACID,
    4-CYANO-, ETHYL ESTER (12); (116460-89-0), **79**, 35

Ethyl p-dimethylaminobenzoate: Benzoic acid, 4-dimethylamino-, ethyl
    ester; (10287-53-3), **76**, 276

9-ETHYL-3,6-DIMETHYLCARBAZOLE: CARBAZOLE, 9-ETHYL-
    3,6-DIMETHYL-; (51545-42-7), **77**, 153

Ethyl 2,2-dimethylpropanoate: Propanoic acid, 2,2-dimethyl-, ethyl ester;

D-Glucose: α-D-Glucopyranose; (492-62-6), **78**, 124

L-(S)-Glyceraldehyde acetonide (2,3-O-Isopropylidene-L-glyceraldehyde):
    1,3-Dioxolane-4-carboxaldehyde, (S)-; (22323-80-4), **75**, 139

Glycine methyl ester; (616-34-20), **76**, 58

Glycine methyl ester hydrochloride: Glycine methyl ester, hydrochloride;
    (5680-79-5), **76**, 66

Glyoxylation of amines with chloral hemiacetal, **79**, 196

Green Chemistry, **76**, 174

GRIGNARD REAGENTS, **76**, 87, 221, 222, 228

Guanidine hydrochloride: Guanidine, monohydrochloride; (50-01-1), **78**, 92

Haloalkenes, as synthons, **76**, 224

HALOBORATION, of 1-alkynes, **75**, 134
    OF ALLENE, **75**, 129

Hanovia mercury lamp, **76**, 276; **77**, 122, 165

Hexachloroethane: Ethane, hexachloro-; (67-72-1), **78**, 83

Hexafluoroisopropyl alcohol: 2-Propanol, 1,1,1,3,3,3-hexafluoro-; (920-66-1),
    76, 151

Hexafluorophosphate salts, vs. perchlorate salts, **77**, 210

2,3,5,6,8,9-Hexahydrodiimidazo[1,2-a:2',1'-c]pyrazine: Diimidazo[1,2-a:2',1'-c]
    pyrazine, 2,3,5,6,8,9-hexahydro-; (180588-23-2), **78**, 73

Hexamethyldisilane, **75**, 155

Hexamethyldisilazane: Silanamine, 1,1,1-trimethyl-N-(trimethylsilyl)-; (999-97-3),
    78, 113

Hexamethylenetetramine: 1,3,5,7-Tetraazatricyclo[3.3.1.1^{3,7}]decane; (100-97-0),
    75, 3

Hexamethylphosphoric triamide, HIGHLY TOXIC: Phosphoric triamide,
    hexamethyl-; (680-31-9), **76**, 201; **78**, 105

Hexylamine: 1-Hexanamine; (111-26-2), **78**, 24

N-HEXYL-2-METHYL-4-METHOXYANILINE, **78**, 23

1-Hexyne; (693-02-7), **76**, 191

Hunsdiecker reaction, **75**, 127

Hydratropic acids, syntheses, **76**, 173

Hydrazine monohydrate, HIGHLY TOXIC. CANCER SUSPECT AGENT: Hydrazine;

from racemic material, **79**, 76

(1S,2S)-N-(2-Hydroxy-1-methyl-2-phenylethyl)-N-methylpropionamide:
Propanamide, N-(2-hydroxy-1-methyl-2-phenylethyl)-N-methyl-
[R-(R*,R*)]- (192060-67-6); [S-(R*,R*)]-; (159213-03-3), **77**, 22

[1S(R)2S]-N-(2-HYDROXY-1-METHYL-2-PHENYLETHYL)-N,2-
DIMETHYLBENZENEPROPIONAMIDE OR (1S,2S)-
PSEUDOEPHEDRINE-(R)-2-METHYLHYDROCINNAMIDE:
BENZENEPROPANAMIDE, N-(2-HYDROXY-1-METHYL-
2-PHENYLETHYL)-N,α-DIMETHYL-, [1S-[1R*(R*),2R*]]-; (159345-08-1);
[1S-[1R*(S*),2R]]-; (159345-06-9), **77**, 23, 30

2-Hydroxy-4-methylpyridine: 2(1H)-Pyridinone, 4-methyl-; (13466-41-6), **78**, 52

2-Hydroxy-5-methylpyridine: 2(1H)-Pyridinone, 5-methyl-; (1003-68-5), **78**, 51

2-Hydroxy-6-methylpyridine: 2(1H)-Pyridinone, 6-methyl-; (3279-76-3), **78**, 52

4-HYDROXY-1,1,1,3,3-PENTAFLUORO-2-HEXANONE HYDRATE:
2,2,4-HEXANETRIOL,1,1,1,3,3-PENTAFLUORO-; (119333-90-3), **76**, 151

3-Hydroxyquinuclidine: 1-Azabicyclo[2.2.2]octan-3-ol; (1619-34-7), **77**, 109

N-Hydroxythiazolethiones, table, **79**, 234

N-Hydroxythiopyridone: 2(1H)-Pyridinethione, 1-hydroxy-; (1121-30-8), **75**, 124

(R)-(+)-2-Hydroxy-1,1,2-triphenyl-1,2-ethanediol: 1,2-Ethanediol,
1,1,2-triphenyl-, (R)-; (95061-246-4), **77**, 45

(R)-(+)-2-HYDROXY-1,2,2-TRIPHENYLETHYL ACETATE [(R)-HYTRA]:
1,2-ETHANEDIOL, 1,1,2-TRIPHENYL-, 2-ACETATE, (R)-; (95061-47-5),
77, 45

(R)- and (S)-HYTRA (2-Hydroxy-1,2,2-triphenylethyl acetate) esters, **77**, 45

Imidazole: 1H-Imidazole; (288-32-4), **75**, 140, **77**, 225

[(2-)-N,O,O'[2,2'-Iminobis[ethanolato]]]-2-butylboron, **76**, 88

Indene: 1H-Indene; (95-13-6), **76**, 47

(1S,2R)-Indene oxide: 6H-Indeno[1,2-b]oxirene, 1a,6a-dihydro-;
(768-22-9), **76**, 47

Indigo test, **76**, 80

Indinavir (Crixivan®), **76**, 52

Indium; (7440-74-6), **77**, 108

Indium (I) iodide: Indium iodide (8); Indium iodide (InI) (9); (13966-94-4), **79**, 59

Indium (III) iodide: Indium iodide (8); Indium iodide (InI$_3$) (9); (13510-35-5), **79**, 59

Indole: 1H-Indole; (120-72-9), **76**, 80

Indoloquinolizines, **76**, 36

Iodine; (7553-56-2), **75**, 69, 100; **76**, 13; **77**, 207; **78**, 105; **79**, 176

4-Iodoacetophenone: Ethanone, 1-(4-iodophenyl)-; (13329-40-3), **77**, 135

o-Iodoaniline: Benzenamine, 2-iodo-; (615-43-0), **78**, 36

4-Iodoanisole: Benzene, 1-iodo-4-methoxy-; (696-62-8), **75**, 61

2-Iodobenzoic acid: Benzoic acid, 2-iodo-; (88-67-5), **77**, 141

2-IODO-2-CYCLOHEXEN-1-ONE: 2-CYCLOHEXEN-1-ONE, 2-IODO-; (33948-36-6), **75**, 69

Iron(III) chloride hexahydrate: Iron chloride, hexahydrate; (10025-77-1), **78**, 249

Irradiation, **76**, 276; **77**, 122, 165; **79**, 165

Irradiation apparatus, **75**, 141

Isobutyraldehyde: Propanal, 2-methyl-; (78-84-2), **77**, 109; **79**, 116

Isobutyronitrile(8): Propanenitrile, 2-methyl- (9); (78-82-0), **79**, 209

Isobutyryl chloride: See 2-Methylpropanoyl chloride, **75**, 118

ISOMERIZATION OF β-ALKYNYL ALLYLIC ALCOHOLS TO FURANS CATALYZED BY SILVER NITRATE ON SILICA GEL, **76**, 263

Isomerization, of a meso-diamine to the dl-isomer, **76**, 25

Isoprene: 1,3-Butadiene, 2-methyl-; (78-79-5), **76**, 25

Isopropenyl acetate: 1-Propen-2-ol, acetate; (108-22-5), **77**, 125

O^4,O^5-Isopropylidene-3,6-anhydro-1-deoxy-1-iodo-D-glucitol, **77**, 91

O^4,O^5-ISOPROPYLIDENE-1,2:3,6-DIANHYDRO-D-GLUCITOL, **77**, 91

2,3-O-Isopropylidene-L-glyceraldehyde: 1,3-Dioxolane-4-carboxaldehyde, 2,2-dimethyl-, L-; (22323-80-4), **75**, 139

2,3-O-Isopropylidene-L-threitol: 1,3-Dioxolane-4,5-dimethanol, 2,2-dimethyl-, (4S-trans)-; (50622-09-8), **76**, 102

2,3-O-Isopropylidene-L-threitol 1,4-bismethanesulfonate, **76**, 101

(3R*,4R*)- and (3R*,4S*)-4-Isopropyl-4-methyl-3-octyl-2-oxetanone, **75**, 119

Isosorbide: D-Glucitol, 1,4:3,6-dianhydro-; (652-67-5), **77**, 91

Isovaleraldehyde: Butanal, 3-methyl-; (590-86-3), **75**, 20

Lithium methoxide: Methanol, lithium salt; (865-34-9), **76**, 58

LITHIUM PENTAFLUOROPROPEN-2-OLATE: 1-PROPEN-2-OL,
1,1,3,3,3-PENTAFLUORO-, LITHIUM SALT; (116019-90-0), **76**, 151

Lithium phenyltrimethoxyborate (in situ), **79**, 85

Lithium triethoxyaluminum hydride: Aluminate (1-), triethoxyhydro-, lithium (I-4);
(17250-30-5), **77**, 30

Lithium wire; (7439-93-2), **76**, 25

2,6-Lutidine: Pyridine, 2,6-dimethyl-; (108-48-5), **77**, 163

Magnesium; (7439-95-4), **75**, 107; **76**, 13, 87, 221; **78**, 104; **79**, 1, 176

Magnesium perchlorate hexahydrate: Perchloric acid, magnesium salt,
hexahydrate; (13446-19-0), **77**, 164

Malononitrile: HIGHLY TOXIC. Propanedinitrile; (109-77-3), **75**, 210

Manganese acetate tetrahydrate: Acetic acid, manganese (2+ salt),
tetrahydrate;
(6156-78-1), **75**, 4

Manganese(II) chloride tetrahydrate, **76**, 242

Manganese(II) chloride: Manganese chloride; (7773-01-5), **76**, 241, 242

Manganese(II) sulfate, **76**, 79

McMurry olefin synthesis, two extensions to, **76**, 145

Meldrum's acid: 1,3-Dioxane-4,6-dione, 2,2-dimethyl-; (2033-24-1), **77**, 114

l-Menthol, **76**, 29

(1R,2S,5R)-(-)-Menthyl (S)-p-toluenesulfinate: Benzenesulfinic acid,
4-methyl-, 5-methyl-2-(1-methylethyl)cyclohexyl ester,
[1R-[1α(S*),2β,5α]]; (188447-91-8), **77**, 51

β-MERCAPTOPROPIONITRILE: PROPANENITRILE, 3-MERCAPTO-;
(1001-58-7), **77**, 186

Meroquinene tert-butyl ester: 4-Piperidineacetic acid, 3-ethenyl-, 1,1-
dimethylethylester, (3R-cis)-; (52346-11-9), **75**, 225

Meroquinene esters, **75**, 231

2-(N-Mesitylenesulfonyl)amino-1-phenyl-1-propanol: Benzenesulfonamide,
N-(2-hydroxy-1-methyl-2-phenylethyl)-2,4,6-trimethyl-, [S-(R*,S*)]- (14);
(187324-62-5), **79**, 109

Mesitylenesulfonyl chloride: 2-Mesitylenesulfonyl chloride (8); Benzenesulfonyl

(R)-(+)- and (S)-(-)-α-Methoxy-α-(trifluoromethyl)phenylacetic acid:
Benzeneacetic acid, α-methoxy-β,β,β-(trifluoromethyl)-, (R)-;
(20445-31-2), (S)-; (17257-71-5), **77**, 37, 72;

(R)-(-)-α-Methoxy-α-(trifluoromethyl)phenylacetyl chloride: Benzeneacetyl
chloride, α–methoxy-β,β,β-(trifluoromethyl)-, (R)-; (20445-33-4), **77**, 7, 40

Methyl acetate: Acetic acid, methyl ester; (79-20-9), **77**, 52

Methyl acetimidate hydrochloride: Acetimidic acid, methyl estyer, hydrochloride
(8); Ethanimidic acid, methyl ester, hydrochloride (9); (14777-27-6),
79, 244

Methyl acrylate: 2-Propenoic acid, methyl ester; (96-33-3), **75**, 106; **77**, 109;
78, 161

Methylamine: Methanamine; (74-89-5), **76**, 24

N-Methylaniline: Benzenamine, N-methyl-; (100-61-8), **78**, 24

(R)-α-METHYLBENZENEPROPANAL: BENZENEPROPANAL,
α-METHYL- (R)-; (42307-9-5), **77**, 29

(R)-α-Methylbenzenepropanoic acid: Benzenepropanoic acid, α-methyl-
(R)-; (14367-67-0), **77**, 35

(R)-β-METHYLBENZENEPROPANOL: BENZENEPROPANOL,
β-METHYL-, (R)-; (77493-96-5), **77**, 31

N-Methylbenzimine: Methanamine, N-(phenylmethylene)-; (622-29-7), **76**, 24

(R)-(+)-α-Methylbenzylamine: Benzenemethanamine, α-methyl-, (R)-;
(3886-69-9), **77**, 35

Methyl 2-(benzylamino)methyl-3-hydroxybutanoate: Butanoic acid, 3-hydroxy-
2-[[(phenylmethyl)amino]methyl]-, methyl ester; (R*,R*)-; (118559-03-8);
(R*,S*)-; (118558-99-9), **75**, 107

(R)-N-(α-Methylbenzyl)-α-methylbenzenepropanamide, **77**, 35

4-Methyl-2,2'-bipyridine: 2,2'-Bipyridine, 4-methyl-; (56100-19-7), **78**, 51

5-METHYL-2,2'-BIPYRIDINE: 2,2'-BIPYRIDINE, 5-METHYL-; (956100-20-0),
78, 51

6-Methyl-2,2'-bipyridine: 2,2'-Bipyridine, 6-methyl-; (56100-22-2), **78**, 51

Methyl (Z)-2-(bromomethyl)-4-methylpent-2-enoate: 2-Pentenoic acid,
2-(bromomethyl)-4-methyl-, methyl ester, (Z)-; (137104-29-3), **77**, 107

3-Methylbutan-2-one; (563-80-4), **75**, 119

2-Methyl-2-butene: 2-Butene, 2-methyl-; (513-35-9), **77**, 34

3-Methyl-3-buten-1-ol: 3-Buten-1-ol, 3-methyl-; (763-32-6), **78**, 203

Methyl (R)-3-(tert-butyldiphenylsilyloxy)-2-methylpropionate: Propanoic acid, [[(1,1-dimethylethyl)diphenylsilyl]oxy]-2-methyl-, methyl ester, (2R)- (13); (153775-90-7), **79**, 59

Methyl tert-butyl ether: Propane, 2-methoxy-2-methyl- (9); (1634-04-4), **75**, 3; **77**, 199

(S)-2-Methyl-CBS-oxazaborolidine: 1H, 3H-Pyrrolo[1,2-c][1,3,2]oxazaborole, tetrahydro-1-methyl-3,3-diphenyl-, (S)- (12); (112022-81-8), **79**, 72

N-METHYL-N-(4-CHLOROPHENYL)ANILINE: BENZENAMINE, 4-CHLORO-N-METHYL-N-PHENYL-; (174307-94-9), **78**, 23

2-METHYLCYCLOHEXANONE: CYCLOHEXANONE, 2-METHYL-; (583-60-8), **76**, 240

Methyl 3-cyclopentene-1-carboxylate: 3-Cyclopentene-1-carboxylic acid, methyl ester; (58101-60-3), **75**, 197

Methyl cyclopropanecarboxylate: Cyclopropanecarboxylic acid, methyl ester; (2868-37-3), **78**, 142

(1'S,2'S)-METHYL-3O,4O-(1',2'-DIMETHOXYCYCLOHEXANE-1',2'-DIYL)-α-D-MANNOPYRANOSIDE: (α-D-MANNOPYRANOSIDE, METHYL 3,4-O-(1,2-DIMETHOXY-1,2-CYCLOHEXANEDIYL)-, [3[S(S)]]-); (163125-35-7), **75**, 170

METHYL 5,5-DIMETHYL-4-OXOHEXANOATE: HEXANOIC ACID, 5,5-DIMETHYL-4-OXO-, METHYL ESTER (9); (34553-32-7), **79**, 146

Methyl 4,4-dimethyl-4-oxopentanoate Pentanoic acid, 4,4-dimethyl-3-oxo-, methyl ester (9); (55107-14-7), **79**, 146

(3R),(4E)-METHYL 3-(DIMETHYLPHENYLSILYL)-4-HEXENOATE: 4-HEXENOIC ACID, 3-(DIMETHYLPHENYLSILYL)-, METHYL ESTER, [R-(E)]-; (136174-52-2), **75**, 78
ee determination, **75**, 83

(3S),(4E)-METHYL 3-(DIMETHYLPHENYLSILYL)-4-HEXENOATE: 4-HEXENOIC ACID, 3-(DIMETHYLPHENYLSILYL)-, METHYL ESTER, [S-(E)]-; (136314-66-4), **75**, 78
ee determination, **75**, 82

4-Methyl-1,3-dioxane: 1,3-Dioxane, 4-methyl-; (1120-97-4), **75**, 177

trans-2-METHYL-2,3-DIPHENYLOXIRANE: OXIRANE, 2-METHYL-2,3-DIPHENYL-, trans-; (23355-99-9), **78**, 225

7-Methylene-8-hexadecyn-6-ol: 8-Hexadecyn-6-ol, 7-methylene-; (170233-66-6), **76**, 264

Methylene iodide, **79**, 146

Methyl formate: Formic acid, methyl ester; (107-31-3), **75**, 171; **79**, 244

Methyl α-D-galactopyranoside: α-D-Galactopyranoside, methyl; (3396-99-4), **77**, 213

Methyl glycinate hydrochloride: Glycine, methyl ester, hydrochloride (8,9); (5680-79-5), **79**, 244

Methyl 3-hydroxy-2-methylenebutanoate: Butanoic acid, 3-hydroxy-2-methylene-, methyl ester; (18020-65-0), **75**, 106

METHYL 3-HYDROXYMETHYL-4-METHYL-2-METHYLENE-PENTANOATE: PENTANOIC ACID, 3-HYDROXY-4-METHYL-2-METHYLENE-, METHYL ESTER; (71385-30-1), **77**, 107

Methyl (R)-(-)-3-hydroxy-2-methylpropionate: Propanoic acid, 3-hydroxy-2-methyl-, methyl ester, (R)- (10); (72657-23-9), **79**, 59

N-Methylimidazole: Imidazole, 1-methyl- (8); 1H-Imidazole, 1-methyl- (9); (616-47-7), **79**, 236

Methyl iodide: Methane, iodo-; (74-88-4), **75**, 19; **78**, 4; **79**, 130

METHYL (S)-2-ISOCYANATO-3-PHENYLPROPANOATE: BENZENEPROPANOIC ACID, α-ISOCYANATO-, METHYL ESTER, (S)-; (40203-94-9), **78**, 220

Methyl (4S)-4,5-O-isopropylidenepent-(2Z)-enoate: 2-Propenoic acid, 3-(2,2-dimethyl-1,3-dioxolan-4-yl)-, methyl ester, [S-(Z)]-; (81703-94-8), **75**, 140

Methyllithium: Lithium, methyl-; (917-54-4), **75**, 99; **76**, 193, 253

Methylmagnesium bromide: Magnesium, bromomethyl-; (75-16-1), **77**, 154; **78**, 44

Methylmagnesium chloride: Magnesium, chloromethyl- (8, 9); (676-58-4), **79**, 19

Methyl α-D-mannopyranoside; (617-04-9), **75**, 171

METHYL METHANETHIOSULFONATE: METHANESULFONOTHIOIC ACID, S-METHYL ESTER; (2949-92-0), **78**, 99

Methyl α-[(methoxyethylidene)amino]acetate: Glycine, N-(1-methoxyethylidene)-, methyl ester (10); (64991-38-4), **79**, 244

METHYL N-(p-METHOXYPHENYL)CARBAMATE: CARBAMIC ACID, N-(4-METHOXYPHENYL)-, METHYL ESTER; (14803-72-6), **78**, 234

METHYL 2,3-O-(6,6'-OCTAHYDRO-6,6'-BI-2H-PYRAN-2,2'-DIYL)-α-D-GALACTOPYRANOSIDE: α-D-GALACTOPYRANOSIDE, METHYL, 2,3-O-(OCTAHYDRO[2,2'-BI-2H-PYRAN]-2,2'-DIYL-, [2(2R,2'R)]-; (144102-32-9), **77**, 212

Methyl Orange: Benzenesulfonic acid, 4-[[4-(dimethylamino)phenyl]azo]-, sodium salt; (547-58-0), **77**, 237

3-Methyl-3-oxetanemethanol: 3-Oxetanemethanol, 3-methyl- (9); (3143-02-0), **79**, 216

METHYL (R)-(+)-β-PHENYLALANATE: BENZENEPROPANOIC ACID, β-AMINO-, (R)-, METHYL ESTER; (37088-67-8), **77**, 50

(R)-2-METHYL-1-PHENYL-3-HEPTANONE: 3-HEPTANONE, 2-METHYL-1-PHENYL-, (R)-; (159213-12-4), **77**, 32

[R-(R*,S*)]-β-METHYL-α-PHENYL-1-PYRROLIDINEETHANOL: 1-PYRROLIDINEETHANOL, β-METHYL-α-PHENYL-, [R-(R*,S*)]-; (127641-25-2), **77**, 12

Methyl (S)-2-phthalimido-4-methylthiobutanoate: 2H-Isoindole-2-acetic acid, 1,3-dihydro-α-[2-(methylthio)ethyl]-1,3-dioxo-, methyl ester, (S)-; (39739-05-4), **76**, 123

METHYL (S)-2-PHTHALIMIDO-4-OXOBUTANOATE: 2H-ISOINDOLE-2-ACETIC ACID, 1,3-DIHYDRO-1,3-DIOXO-α-(2-OXOETHYL)-, METHYL ESTER, (S)-; (137278-36-5), **76**, 123

2-Methyl-2-propanol: 2-Propanol, 2-methyl-; (75-65-0), **75**, 225; **77**, 34

2-Methylpropanoyl chloride: Propanoyl chloride, 2-methyl-; (79-30-1), **75**, 118

2-METHYL-4H-PYRAN-4-ONE: 4H-PYRAN-4-ONE, 2-METHYL-; (5848-33-9), **77**, 115

4-Methyl-2-pyridyl triflate: Methanesulfonic acid, trifluoro-, 4-methyl-2-pyridinyl ester; (179260-78-7), **78**, 52

6-Methyl-2-pyridyl triflate: Methanesulfonic acid, trifluoro-, 6-methyl-2-pyridinyl ester; (154447-04-8), **78**, 52

1-Methyl-2-pyrrolidinone: 2-Pyrrolidinone, 1-methyl-; (872-50-4), **76**, 240; **77**, 135; **79**, 35

Methyl serinate hydrochloride, **77**, 65

trans-α-Methylstilbene: Benzene, 1,1'-(1-methyl-1,2-ethenediyl)bis-, (E)-; (833-81-8), **78**, 225

α-Methylstyrene: Benzene, (1-methylethenyl)-; (98-83-9), **76**, 159

(S$_S$,R)-(+)-Methyl N-(p-toluenesulfinyl)-3-amino-3-phenylpropanoate: Benzenepropanoic acid, β-[[(4-methylphenyl)sulfinyl]amino]-, methyl ester, [S-(R*,S*)]-; (158009-86-0), **77**, 52

3-Methyl-3-(toluenesulfonyloxymethyl)oxetane: 3-Oxetanemethanol, 3-methyl-, 4-methylbenzenesulfonate (11); (99314-44-0), **79**, 216

Methyl triflate: Methyl trifluoromethanesulfonate: Methanesulfonic acid, trifluoro-, methyl ester; (333-27-7), **77**, 81

5-Methyl-2-(trifluoromethanesulfonyl)oxypyridine: Methanesulfonic acid, trifluoro-, 5-methyl-2-pyridinyl ester; (154447-03-7), **78**, 52

Methyl (triphenylphosphoranylidene)acetate: Propanoic acid, 2-(triphenylphosphoranylidene)-, methyl ester; (2605-67-6), **75**, 139

Methyl vinyl ketone: 3-Buten-2-one; (78-94-4), **78**, 249

Mineral salt bath (MSB), **76**, 77

Mitsunobu displacement, **76**, 185

4 Å Molecular sieves: Zeolites, 4 Å; (70955-01-0), **75**, 12, 189

Molybdenum hexacarbonyl: Molybdenum carbonyl (8); Molbdenum carbonyl, (OC-6-11) (9); (13939-06-5), **79**, 27

MONOALKYLATION, REGIOSELECTIVE, OF KETONES, VIA MANGANESE ENOLATES, **76**, 239

MONO-C-METHYLATION OF ARYLACETONITRILES AND METHYL ARYLACETATES BY DIMETHYL CARBONATE, **76**, 169

Morpholine, **79**, 52

Mosher esters, **77**, 7, 37, 71

2-NAPHTHYLMAGNESIUM BROMIDE: MAGNESIUM, BROMO-2-NAPHTHALENYL-; (21473-01-8), **76**, 13

Nickel acetylacetonate: Nickel, bis(2,4-pentanedionato-) (8); Nickel, bis(2,4-pentane dionato-O,O')- (sp-4-1)- (9); (3264-82-2), **79**, 35

Nitric acid; (7697-37-2), **75**, 90

Nitriles, into tertiary amines, **76**, 133

Nitrilotriacetic acid, CANCER SUSPECT AGENT: Glycine, N-bis(carboxymethyl)-; (139-13-9), **76**, 78

Nitrobenzene: Benzene, nitro-; (98-95-3), **77**, 2

o-Nitrobenzenesulfinic acid: Benzenesulfinic acid, 2-nitro-; (13165-79-2), **76**, 185

o-Nitrobenzenesulfonyl chloride: Benzenesulfonyl chloride, 2-nitro-; (1694-92-4), **76**, 183; **79**, 186

o-Nitrobenzenesulfonyl hydrazide: Benzenesulfonic acid, 2-nitro-, hydrazide; (5906-99-0), **76**, 180, 183

2-Nitroethanol: Ethanol, 2-nitro-; (625-48-9), **78**, 16

Nitrogen oxides, **75**, 90

(2S,3S)-2-NITRO-5-PHENYL-1,3-PENTANEDIOL, **78**, 14

3-NITROPROPANAL: PROPANAL, 3-NITRO-; (58657-26-4), **77**, 236

Oxone: Peroxymonosulfuric acid, monopotassium salt, mixt. with dipotassium sulfate and potassium hydrogen sulfate; (37222-66-5), **78**, 225

Ozone; (10028-15-6), **78**, 254

Palladium, 10% on carbon, **75**, 226; **77**, 135

Palladium acetate: Acetic acid, palladium (2+) salt; (3375-31-3), **75**, 53, 61; **78**, 3, 24, 36; **79**, 27, 52, 72

Palladium(II) bis(benzonitrile)dichloride: Palladium, bis(benzonitrile)dichloro-; (14220-64-5), **75**, 70

Palladium (II) chloride bisacetonitrile: Palladium, bis(acetonitrile)dichloro-; (14592-56-4), **77**, 213

PALLADIUM CATALYST, LIGANDLESS, **75**, 61

20% Palladium hydroxide on carbon, **75**, 21

Paraformaldehyde; (30525-89-4), **78**, 169

Paraldehyde: 1,3,5-Trioxane, 2,4,6-trimethyl-; (123-53-7), **78**, 124

Parr shaker, **75**, 24, 226

PENTA-1,2-DIEN-4-ONE: 3,4-PENTADIEN-2-ONE; (2200-53-5), **78**, 135

Pentaerythritol: 1,3-Propanediol, 2,2-bis(hydroxymethyl)-; (115-77-5), **75**, 89

Pentaerythrityl tetrachloride: Propane, 1,3-dichloro-2,2-bis(chloromethyl)-; (3228-99-7), **75**, 89

Pentaerythrityl trichlorohydrin: 1-Propanol, 3-chloro-2,2-bis(chloromethyl)-; (813-99-0), **75**, 89

2,4-Pentanedione; (123-54-6), **77**, 2

2-Pentanol; (6032-29-7), **76**, 255

Pentan-3-one; (96-22-0), **79**, 251

2-PENTYL-3-METHYL-5-HEPTYLFURAN: FURAN, 5-HEPTYL-3-METHYL-2-PENTYL-; (170233-67-7), **76**, 263

Perchlorate salts, **77**, 210

2-Perfluorohexyl-1-iodoethane: Octane, 1,1,1,2,2,3,3,4,4,5,5,6,6-tridecafluoro-8-iodo- (9); (2043-57-4), **79**, 1

cis-4a(S),8a(R)-PERHYDRO-6(2H)-ISOQUINOLINONES, **75**, 223

pH 7 Buffer, **76**, 253

Phase transfer catalysis, **75**, 103; **76**, 273

Potassium hydride; (7693-26-7), **75**, 20; **79**, 130

Potassium methyl α-[(methoxyethylidene)amino]-β-hydroxyacrylate: Propanoic acid, 2-[(1-methoxyethylidene)amino]-3-oxo-, methyl ester, ion(1-), potassium (11); (105205-35-5), **79**, 244

Potassium tetrafluoroborate: Borate(1-), tetrafluoro-, potassium (8,9); (14075-53-7), **79**, 236

Preculture preparation, **76**, 77

4a,9a-Propano-4H-cyclopenta[5,6]pyrano[2,3-d]-1,3-dioxin-6,12(5H)-dione, **75**, 191

Propanoic acid: See Propionic acid, **75**, 80

1-Propanol; (71-23-8), **75**, 53

Propargyl alcohol: 2-Propyn-1-ol; (107-19-7), **76**, 178, 200

2-Propargyloxytetrahydropyran: 2H-Pyran, tetrahydro-2-(2-propynyloxy)-; (6089-04-9), **76**, 178

[1.1.1]PROPELLANE: TRICYCLO[1.1.1.0^{1,3}]PENTANE; (35634-10-7), **75**, 98; **77**, 249

(Z)-4-(2-PROPENYL)-3-OCTEN-1-OL: 3-OCTEN-1-OL, 4-(2-PROPENYL)-, (Z)- (12); (119528-99-3), **79**, 11

Propionaldehyde: Propanal; (123-38-6), **76**, 152

Propionic acid: Propanoic acid; (79-09-4), **75**, 80

Propionic anhydride: Propanoic acid, anhydride; (123-62-6), **77**, 22

Propyl alcohol: See 1-Propanol, **75**, 53

1-PROPYNYLLITHIUM: LITHIUM, 1-PROPYNYL-; (4529-04-8), **76**, 214

Protected vicinal amino alcohols, synthesis of, **77**, 78

Protection, for diequatorial vicinal diols, **77**, 218
    of trans-hydroxyl groups in sugars, **75**, 173
        table, **75**, 174
    of primary and secondary amines, **75**, 167, 232; **79**, 186
    selective, of 1,3-diols, **75**, 177
        table, **75**, 182
    selective, of primary alcohols, **75**, 186

Protective groups:
    Boc-, **76**, 72
    2-cyanoethyl, **77**, 193
    1-ethoxyethyl, removal of, **76**, 207
    in phosphoramidite syntheses, **76**, 273
    2-nitrophenylsulfonyl, **79**, 186

endo-TRICYCLO[3.2.1.0^{2,4}]OCT-6-ENE:TRICYCLO[3.2.1.0^{2,4}]OCT-6-ENE, (1α, 2α, 4α, 5α,)-; (3635-94-7), **77**, 255

TRICYCLO[1.1.1.0^{1,3}]PENTANE: See [1.1.1]PROPELLANE, **75**, 98

Triethylamine: Ethanamine, N,N-diethyl-; (121-44-8), **75**, 45, 108; **76**, 112, 123, 202, 264; **77**, 3, 65, 80, 163, 200; **78**, 4, 136, 179; **79**, 27, 109, 116, 154, 165, 186, 251

Triethylamine hydrobromide: Ethanamine, N,N-diethyl-, hydrobromide; (636-70-4), **78**, 136

Triethylamine trishydrofluoride: Ethanamine, N,N-diethyl-, trishydrofluoride; (73602-61-6), **76**, 159

Triethylenetetramine: 1,2-Ethanediamine, N,N'-bis(2-aminoethyl)-; (112-24-3), **78**, 74

Triethyl orthoformate: Ethane, 1,1',1"-[methylidynetris(oxy)]tris-; (122-51-0), **77**, 241

Triethyl phosphite: Phosphorous acid, triethyl ester; (122-52-1), **78**, 170

Triethyl phosphonoacetate: Acetic acid, (diethoxyphosphinyl)-, ethyl ester; (867-13-0), **78**, 125

Triflic anhydride: Methanesulfonic acid, trifluoro-, anhydride; (358-23-6), **76**, 81; **79**, 43

Trifluoroacetic acid: Acetic acid, trifluoro-; (76-05-1), **75**, 130; **76**, 207, 288; **77**, 30, 53; **79**, 154

Trifluoroacetic anhydride: Acetic acid, trifluoro-, anhydride; (407-25-0), **76**, 95, 190, 193

1,1,1-TRIFLUORO-2-ETHOXY-2,3-EPOXY-5-PHENYLPENTANE: OXIRANE, 2-ETHOXY-3-(2-PHENYLETHYL)-2-(TRIFLUOROMETHYL)-, cis-(±)-; (141937-91-9), **75**, 153

(Z)-1,1,1-TRIFLUORO-2-ETHOXY-5-PHENYL-2-PENTENE: BENZENE, (4-ETHOXY-5,5,5-TRIFLUORO-3-PENTENYL)-, (Z)-; (141708-71-6), **75**, 153

Trifluoromethanesulfonic acid: Methanesulfonic acid, trifluoro-; (1493-13-6), **76**, 294; **78**, 105, 178; **79**, 103

Trifluoromethanesulfonic anhydride: Methanesulfonic acid, trifluoro-, anhydride; (358-23-6), **78**, 2, 27, 52, 91

3-(Trifluoromethyl)benzoyl chloride: Benzoyl chloride, 3-(trifluoromethyl)-; (2251-65-2), **77**, 163

1-Triisopropylsilyloxy-1-azidocyclohexane: Silane, [(1-azidocyclohexyl)oxy]tris (1-methylethyl)- (13); (172090-42-5), **79**, 165

1-Triisopropylsilyloxycyclohexene: Silane, (1-cyclohexen-1-yloxy)tris(1-

77, 176

Tris(2-perfluorohexylethyl)phenyltin: Stannane, tris(3,3,4,4,5,5,6,6,7,7,8,8,8-tridecafluorooctyl)- (13); (175354-31-1), **79**, 1

TRIS(2-PERFLUOROHEXYLETHYL)TIN HYDRIDE: STANNANE, PHENYLTRIS(3,3,4,4,5,5,6,6,7,7,8,8,8-TRIDECAFLUOROOCTYL)- (13); **79**, 1

Trisubstituted alkenes, stereoselective synthesis of, **79**, 15

Tropylium tetrafluoroborate: See Cycloheptatrienylium tetrafluoroborate, **75**, 210

Tryptones (Bacteriological), See: Peptones, Bacteriological; (73049-73-7), **76**, 80

Ultrasonic cleaning bath, **77**, 81

Undecenoic acid: 10-Undecenoic acid ; (112-38-9), **75**, 124

N-(10-Undecenoyloxy)pyridine-2-thione: 2(1H)-Pyridinethione, 1-[(1-oxo-10-undecenyl)oxy]-; (114050-28-1), **75**, 124

Vilsmeier-Haack reagent, **75**, 167

Vinyl acetate: Acetic acid ethenyl ester; (108-05-4), **75**, 79

N-Vinylation, **75**, 45

Vinyllithium: Lithium, ethenyl-; (917-57-7), **76**, 190, 193

N-Vinyl-2-pyrrolin-2-one: 2-Pyrrolidinone, 1-ethenyl-; (88-12-0), **75**, 215

Vinyltriethoxysilane: Silane, triethoxyvinyl- (8); Silane, ethenyltriethoxy- (9); (78-08-0), **79**, 52

Vinyltrimethylsilane: Silane, ethenyltrimethyl-; (754-05-2), **75**, 161

VITAMIN D_2: 9,10-SECOERGOSTA-5,7,10(19),22-TETRAEN-3-OL, (3β)-; (50-14-6), **76**, 275

Vitamin D_2 3,5-dinitrobenzoate: Ergocalciferol, 3,5-dinitrobenzoate; (4712-11-2), **76**, 276

Vortex mixer, **75**, 2

WITTIG OLEFINATION, OF PERFLUORO ALKYL CARBOXYLIC ESTERS, **75**, 153

Zeolites, 4 Å: See 4 Å Molecular sieves, **75**, 189

Zinc; (7440-66-6), **76**, 24, 143, 252; **78**, 99